T0329918

MEDIA TECHNOLOGIES FOR WORK AND PLAY IN EAST ASIA

Critical Perspectives on Japan
and the Two Koreas

Edited by
Micky Lee and Peichi Chung

BRISTOL
UNIVERSITY
PRESS

First published in Great Britain in 2021 by

Bristol University Press
University of Bristol
1-9 Old Park Hill
Bristol
BS2 8BB
UK
t: +44 (0)117 954 5940
e: bup-info@bristol.ac.uk

Details of international sales and distribution partners are available at bristoluniversitypress.co.uk

British Library Cataloguing in Publication Data
A catalogue record for this book is available from the British Library

ISBN 978-1-5292-1336-2 hardcover
ISBN 978-1-5292-1337-9 ePub
ISBN 978-1-5292-1338-6 ePdf

Cover design: Liam Roberts
Front cover image: Unsplash layMbSJ3YOE

Bristol University Press uses environmentally responsible print partners.

Printed in Great Britain by CPI Group (UK) Ltd, Croydon, CR0 4YY

Contents

List of Figures and Tables

Tables

Figures

Notes on Contributors and Editors

Keiji Amano is a Professor of Cultural Economy at Seijoh University, Japan where he leads the e-Sports Management Programme. He received his Master's in International Relations from the Graduate School of International Relations, Ritsumeikan University. He contributes a socioeconomic perspective to a trans-disciplinary research project on the Japanese 'Galapagos' game Pachinko with Geoffrey Rockwell.

Keung Yoon Bae is a PhD candidate at Harvard University's East Asian Languages and Civilizations programme, focusing on East Asian film and media. Her dissertation is on film regulation under the Japanese Empire, though her research interests extend to video games and esports. She is an Academy of Korean Studies (AKS) Postdoctoral Research Scholar at Columbia University in 2020–21.

Shawn Bender is an Associate Professor of East Asian Studies and Anthropology at Dickinson College. His current research examines the use of Japanese robotics technologies in the care of dementia and disability worldwide. His first book, *Taiko Boom: Japanese Drumming in Place and Motion* (University of California Press, 2012), focuses on the emergence and popularisation of taiko drumming groups in Japan. His other publications have appeared in the *Journal of Asian Studies* and *Social Science Japan Journal*.

Peichi Chung is an Associate Professor in the Department of Cultural and Religious Studies at the Chinese University of Hong Kong. Her teaching interests include digital culture, emerging immersive technology, and cultural policy. She has published book chapters and journal articles on comparative film and video game industries in the Asia Pacific.

Patrick W. Galbraith is a Lecturer at Senshu University in Tokyo. He is the author of *Otaku and the Struggle for Imagination in Japan* (Duke University Press, 2019), *The Moe Manifesto: An Insider's Look at the Worlds of Manga, Anime, and Gaming* (Tuttle, 2014), co-author of *AKB48* (Bloomsbury, 2019), and co-editor of *Idols and Celebrity in Japanese Media Culture* (Palgrave, 2012).

Ana Gascón Marcén is an Associate Professor at the Faculty of Law of the University of Zaragoza. Her main research topics are law and ICT (information and communication technologies) and Japanese law. Some of her publications are 'La nueva protección de datos personales: Una mirada a Japón desde Europa' in *Los Derechos Individuales en el Ordenamiento Japonés (*Aranzadi, 2016), 'The extraterritorial application of European Union Data Protection Law' in the *Spanish Yearbook of International Law* (Brill, 2019) and 'The regulation of the flow of personal data between the European Union and the United Kingdom after Brexit' (Cuadernos de derecho transnacional, 2020).

Kyooeun Jang received her BA in Comparative Literature and Culture with a minor in Political Science from Yonsei University. Having worked in digital advertising and policy operations at Facebook, Kyooeun pursued her research interests including digital anthropology, digitally mediated representations of gender, and science and technology studies at Harvard University's Regional Studies East Asia programme. She is currently a PhD student at the Annenberg School of Communication, University of Pennsylvania.

Dal Yong Jin is Distinguished SFU Professor at Simon Fraser University. His major research and teaching interests are on globalisation and media, transnational cultural studies, digital platforms and digital gaming, and the political economy of media and culture. He has published numerous books, including *Korea's Online Gaming Empire* (MIT Press, 2010), *De-convergence of Global Media Industries* (Routledge, 2015)*, Digital Platforms, Imperialism and Political Culture* (Routledge, 2015), *New Korean Wave: Transnational Cultural Power in the Age of Social Media* (University of Illinois Press, 2016), and *Smartland Korea: Mobile Communication, Culture and Society* (University of Michigan Press, 2017). He is the founding book series editor of *Routledge Research in Digital Media and Culture in Asia*.

Micky Lee is Professor of Media Studies at Suffolk University, Boston. She has published in the areas of feminist political economy, and information, technologies, and finance. Her latest books are *Understanding the Business of Global Media in the Digital Age* (co-authored with Dal Yong Jin, Routledge, 2018), *Bubbles and Machines: Gender, Information, and Financial Crises* (University of Westminster Press, 2019), and *Alphabet: The Becoming of Google* (Routledge, 2019).

Geoffrey Rockwell is a Professor of Philosophy and Humanities Computing at the University of Alberta. He has published on video games, pachinko, textual visualisation, text analysis, and computing in the humanities, including a co-authored book *Hermeneutica: Computer-Assisted Interpretation in the Humanities* (MIT Press, 2016). He is a co-developer of Voyant Tools, a suite of text analysis tools for which he and his co-developer were awarded the CSDH/SCHN 2017 Outstanding Contribution Award. Rockwell also leads the TAPoR project documenting text tools. He is currently the Director of the Kule Institute for Advanced Study and blogs at theoreti.ca.

Elizabeth Shim is United Press International's Chief Asia Writer and co-author of *Korean War in Color* (Seoul Selection, 2010). Shim has been recognised for excellence in reporting by the Center for Strategic and International Studies and is a 2019 recipient of a Korean Collections Consortium of North America grant. She is a graduate of Wellesley College and was a departmental fellow at New York University's Arthur L. Carter Journalism Institute, where she earned her Master's degree in Global Journalism and East Asian Studies. Shim previously reported for the Associated Press and her articles have appeared in the *South China Morning Post*.

Deirdre Sneep did her undergraduate studies in Japanese Studies and her PhD in Urban Theory at the University of Duisburg-Essen. During her studies as well as her subsequent academic career, she has always been fascinated with the relationship between Japanese society and technology, in particular the mobile phone. She is currently working on finding a new research project in the area of digitalisation of the Japanese city.

Maria Toyoda is the Dean of the College of Arts and Sciences and Professor of Government at Suffolk University, Boston. She received her PhD from Georgetown University. Her research interests include financial liberalisation mechanisms, reform of government financial institutions, development banking and foreign aid, the political economy of finance, East Asian economy and politics, foreign aid, and

Japanese politics. Some of her publications have appeared in *Capital Controls* (Edward Elgar, 2015), *Education about Asia*, and *Journal of Asian Studies*.

Weiqi Zhang is an Assistant Professor of Political Science and Legal Studies at Suffolk University, Boston. His research focus on political and economic liberalisation in closed and formerly closed countries, and international relations in Asia. His recent publications focuses on political liberalisation in North Korea and China.

Foreword 1: Media for Work and Play in a Pandemic World

Maria Toyoda

There is a fair chance that readers will react to this volume in ways that the editors and authors probably did not anticipate when they first began working on this project. As I write this foreword, most of us are several weeks into a stay-at-home order issued by leaders across the globe in reaction to the COVID-19 pandemic. Institutions of higher learning, schools, businesses, governments, non-profits, and other organisations have scrambled to move day-to-day operations, if they are able, to various online platforms. Businesses that rely exclusively on brick-and-mortar outlets with face-to-face interaction are struggling to survive and as each day passes there is a growing sense that the world will never be the same.

That both work and play are being rapidly redefined through our technology-mediated interactions underscores the relevancy of this book's themes. While the authors address social transformation through the acceptance and rejection of media technology in East Asian societies, these themes also speak to the global phenomenon of technology diffusion in this period of pandemic and social distancing.

These chapters offer context to ground discussion about media technology utilisation both before and after COVID-19. The idea of place, for instance, has taken on new meanings. To be sure, physical space increasingly competes with virtual, augmented, imagined, ephemeral space for our attention and money.

What is the shape of that competition to come? The authors here contextualise global, regional, and local spaces that determine markets for media content and devices. Several of them explicitly make note of the new social spaces and subcultures that media technology enables and creates. An unexpected characteristic of these new spaces is how non-universal they are, with the occasional exception such as *Pokémon GO* (even in this case, user experiences depend on place-specific, augmented, physical space). Markets for social platforms appear to respect boundaries of all kinds, while also creating and enabling non-place-specific micro-populations of enthusiasts, *otaku*, and dependent users to emerge. They empower individuals to overcome disability, anxiety, and loneliness. Yet they also amplify social dynamics such as inequality, inequity, political and social polarisation, mob mentality, and exclusion.

Though this volume focuses on East Asian societies, the insights can be applied and extended to other societies. A significant share of new media technologies and content originates in East Asia and finds explosive popularity around the globe. But as technology diffuses, the contexts change and local cultures become part of the user experience. East Asia itself contains a myriad of user cultures mediated by national and subnational contexts.

As the authors of this volume suggest, the nature of technology-mediated work and play is fluid, fragmented, politically contested, and richly contextualised. Moreover, new media technologies blur the boundaries between the private and public, the local and global, the real and virtual, and the distinction between work and play. Our enforced quarantines and isolation have deepened our dependence on media technology, accelerating the social transformations that were well under way before the appearance of the novel coronavirus. This is all the more reason to pay attention to the insights developed in these case studies.

Foreword 2: The Development of Information and Communication Technologies in South Korea after World War II

Dal Yong Jin

The Republic of Korea (hereafter South Korea) has been continuing to develop its information and communication technologies (ICTs) since the 1990s. South Korea was in economic hardship after the Japanese colonial period (1910–45) and the Korean War (1950–3). During this period, South Korea's economy worsened, with the widespread destruction of industries and infrastructure. South Korea had no choice but to receive foreign aid. The US contributed the majority of aid because South Korea was a strategically important country in East Asia for US national interests. The South Korean government, under US direction, was able to establish a distinctive economic growth, while transforming its industrial structure to favour both heavy industry (shipbuilding) and light industry (electronics) from the 1960s to 1970s.

However, it was not until the 1990s that South Korea developed its new economic model that focuses on ICTs. In the late 1990s, South Korea experienced the worst economic crisis, known as the International Monetary Fund (IMF) crisis. Post-crisis, the government and businesses shifted their focus to ICTs as a new venture. Although they already recognised the importance of ICTs to the economy, the shift further emphasised that an ICT economy would resuscitate the domestic economy as well as help the country compete on the international market. Information and telecommunications equipment have produced significant changes in South Korea's industrial and export structures in the early twenty-first century. Since 2001, information and telecommunications equipment exports have surpassed semiconductors due to a worldwide recession in the semiconductor business as well as the rapid expansion of the Chinese telecommunications market. During that period, South Korean smartphone manufacturers, including Samsung and LG, solidified their footing in the global market.

The South Korean government and businesses, however, recognised that it is not enough to compete with developed countries such as the US and Japan. Therefore, there has been a push for the rapid deployment of Internet and broadband services. As IT became a key factor in international competitiveness, South Korea promoted the construction and advancement of Internet-related infrastructure for sustainable economic growth (Jin, 2011). South Korean government has constructed an advanced information infrastructure consisting of communications networks, Internet services, application software, computers, and information products and services (Lee et al, 2003). As a result, South Korea has especially advanced digital games, smartphones, digital platforms (like Naver and Daum), and instant mobile messengers (such as KakaoTalk and LINE).

South Korea has developed unique relationships with neighbouring countries, including Japan, China, and North Korea. Most of all, Japan and South Korea have competed against each other in the global IT markets, including semiconductors, television monitors, and handsets. Later, South Korea has closely worked with Japan in the fields of new digital media technologies. For example, LINE is a South Korean messaging system targeted at the Japanese market. LINE is operated by Line Corporation, a subsidiary of the South Korean Internet search engine company, Naver. While Kakao, a competing messaging system, dominated the South Korean market, Naver was expanding their messaging application by targeting foreign messaging markets which have not been developed yet. To this end, Naver released its messaging application LINE to the Japanese messaging market in 2011.

However, South Korea has not developed tangible partnerships with China and North Korea. The level of media technologies in North Korea is not comparable to South Korea's; therefore, South Korean ICT companies do not work with North Korean companies, with the exception of a handful of subsidiaries. South Korean ICT companies, including Samsung and digital game companies, have worked with China. However, due to cultural and political reasons and copyright infringements, several South Korean ICT companies have pulled their subsidiaries out of China and moved them to other countries. For example, in 2011, when Samsung's smartphones were the most popular in the world, its two factories in Huizhou and Tianjin produced and exported 70.1 million and 55.6 million mobile phones, respectively. However, due to rising labour and rental costs, high taxes, and an economic slowdown in China (He, 2019), Samsung shifted its production from China to Southeast Asia, India, and Africa. In the realm of online gaming, South Korea had several subsidiary companies in China. However, China has also rapidly increased its capacity to overtake South Korean companies and become a dominant force in online gaming, which means that China and South Korea are now competing rather than collaborating in the global ICT markets.

References

He, H. (2019) 'Why Samsung's last China smartphone factory closed?', *South China Morning Post*, 15 June, Available from: https://www.scmp.com/economy/china-economy/article/3014564/samsungs-last-china-smartphone-factory-closing-raising [Accessed 22 March 2020].

Jin, D.Y. (2011) *Hands On/Hands Off: The Korean State and the Market Liberalization of the Communication Industry*, New York: Hampton Press.

Lee, H.J., O'Keefe, R.M., and Yun, K.L. (2003) 'The growth of broadband and electronic commerce in South Korea: Contributing factors', *The Information Society*, 19(1): 81–93.

Introduction

Micky Lee and Peichi Chung

We will begin the introduction by looking at esports as a technology for work and play in East Asia. Esports is a good example of how the concepts of work and play are rapidly changing in the region. It also problematises the boundary between work and play, a central question addressed in this volume.

Video gaming has been seen as a leisure activity for children and young people since the 1980s (Kline and de Peuter, 2002). Video gaming was and still is considered juvenile entertainment that will lose its appeal when gamers become working adults (Nichols, 2014). Esports has challenged this assumption in at least two ways: First, gamers are no longer amateurs but professionals trained by the government and industry. They no longer play with friends but compete with each other in organised teams on the international stage. In this sense, esports professionals share similarities with professional athletes: They represent their teams and countries to show off technical skills and sporting spirit. Second, significant investment goes into organising the teams and lobbying for official recognition from sports organisations such as the Olympics. Playing video games is no longer just a pastime undertaken in a private space, but a public spectacle that ignites a sense of community and national belonging. In the Asian context, this technology for work and play has created new meanings of digital work in a state-planned economy.

A recent example that illustrates how esports challenges some assumptions about gaming is the case of it being a demonstration sport in the 2018 Asian Games held in Indonesia. It may not be a surprise that Asia is the pioneer in recognising video gaming as a sport because the region is also a trailblazer in esports game development. Various esports organisations in Asia are setting up gaming rules and training teams for regional and international competition.

However, what is meant by Asia as a 'region'? Scholars have pondered this question. For example, Iwabuchi (2002) believes that Japan had long perceived itself as inferior to the West. Prior to World War II, Japan aimed to emulate the West while seeking domination of its neighbours. It was only in the 1990s that Japan sought to "return" to Asia through market expansion when the region's economy took off. In this sense, Asia is a market for transcultural products. Ge (2010) argues that Asia has been seen as an ideological concept, the opposite of Europe. Historically, the discussion about Asia has borne no direct relation to any geographical considerations of the region. Both scholars share an opinion that Asia is a concept that arose with the East. In addition, Asian countries themselves are still in reckoning with what it is.

Even though countries in Asia appear to see it as a collective, as exemplified by the Asian Games that began after World War II, it has to be acknowledged that the region is as geographically vast as it is culturally diverse. Therefore, we need to first problematise 'Asia' as both a geographical area and a concept in two ways. First, the region is hardly homogeneous in technology development and adaptation. The Asian Games is a good illustration of this uneven development of technology. Among the 18 countries that sent teams to compete in the esports demonstration game, most of the top medallists came from East Asian countries such as China, South Korea, Japan, and Taiwan. In addition, two of the games played in the Asian Games were developed in East Asia: *Area of Valor* developed by China-based Tencent and *Pro Evolution Soccer* created by Japan-based Konami. The disparity between East Asia and the rest of Asia in the areas of competitive gaming and game development shows that Asia cannot be understood as a monolithic entity in the development and consumption of work and play technologies.

Even though Asia is not a monolithic region, 'Asia' is sometimes imagined to be a fixed concept that serves as a negation of the West (Ching, 2000; J.B. Choi, 2010; Chow, 2010; Ge, 2010; Otmazgin, 2013; Hjorth and Khoo, 2016). In other words, Asia is seen as a static entity that has some stable characteristics that the West does not share. Not only does the West see Asia as different from itself, but Asians also see themselves as different from Western beings. For example, to counter the perception that video gaming is violent, some esports representatives from Asia assert that Asian players are inherently different from Western players. The President of the International

Olympics Committee, for example, opposes including esports at the international level despite its success in the Asian Games. He contends that video games promote violence and discrimination, not friendship and solidarity as in other sports (Kaser, 2018). Esports representatives from Asia countered that Asian gamers, unlike their Western counterparts, are not violent, so it should not be a problem (Chakraborty, 2018). This comment implied that unlike Western gamers who enjoy playing games as individual home entertainment, gaming in Asia is often about collective play with friends, online sociality, and national pride. Mass culture is believed to have formed a unique 'Asianism' since World War II (Ching, 2000). Asian gamers believe that they are too peaceful to emulate the shooting violence in games in real life. Asian gamers believe that playing esports games do not emulate the shooting acts in games into real life. Asian gamers play games to enjoy the moment when their skills are levelled as they compete with other players in multi-player online arena (MOBA) games. (See Chapter 2 for a discussion of Japanese game developers making the same argument about sexual acts in PC games.)

The example of the Asian Games shows the tension between work and play technologies as well as that between Asian and non-Asian cultures. Asian countries that are influenced by Confucianism selected specific values to advance their post-World War II modernisation (Tu, 2000; Chen, 2018). Governments and corporations ask workers to exhaust their labour to rebuild the nations. Work ethics is then used to justify why these nations became wealthy. Play is devalued when work is tied to economic success. However, digital technologies have blurred the boundary between work and play as Asian cultures are increasingly globalised. Nonetheless, Asian players see themselves as exceptions who somehow work and play differently from the West. How do we explain these two tensions? Do they mean that work will become more like play in Asia even though the way that Asians work and play will still be seen as different from the West?

These two tensions need to be illuminated from a macro perspective, such as a political economy of communication view. From this perspective, Asian cultures are certainly not an anomaly in an increasingly globalised world, because they can be commercialised and commodified like any culture that appeals to domestic and international markets. Many parallels can be drawn between the commercialisation of esports and that of professional sports. Not only is esports a potential national sport, but it also has huge global market potential. The 2018 Global Esports Market Report stated that esports reaches 165 million enthusiasts worldwide. Esports revenue grew 38 per cent from 2017 to 2018, a market growth of US$906 million (Newzoo, 2018). Like professional athletes in popular global sports such as football (known as soccer in the US), some esports athletes are well-known celebrities among their fans. South Korea's top professional gamer, Sang-Hyeok Lee, whose screen ID is Faker, once broke the Twitch record of the highest number of concurrent users by attracting 245,100 viewers on the first day of his streaming (Chalk, 2017). Like other professional sports, esports competition is a public spectacle attended by supporters in an arena, and watched by television and social media audiences at home. Esports fans nowadays can watch live broadcasts on various media channels such as Twitch, YouTube, MLG.tv, ESPN, and TBS. As in other professional sports, big money is involved in the form of brand sponsorship and advertising. In 2018, the industry successfully attracted sponsorships from global brands such as Coca Cola, Red Bull, Adidas, and McDonald's.

Like established professional sports, esports also suffer from problems such as a concentrated industry and income inequality. A concentrated industry means there is a concentration of profit. While there are few team owners and many players, the former reap most of the profits and leave little to the latter. The industry is also increasingly vertically integrated, with fewer and fewer companies owning more and more of the industry: from team ownership, game development, to venue construction. In addition, the top players also earn much more than the average players. The inequalities in the esports industry show that play technologies are not just childish pastimes, but organised businesses that are not too different from other established businesses such as media, entertainment, and sports. Besides, esports businesses in Asia follow the global business trends of concentration and consolidation. From a macro perspective, it can be argued that technologies for work and play in Asia are not that different from the rest of the world. However, if this is the case, then why does this volume have a regional focus?

Why the Northeast?

This edited book does not look at the entire region of Asia, but focuses on Japan and the two Koreas. Northeast Asia illustrates three points developed in this book well. First, even though these three countries shared Confucian values in political governance and social ethics prior to Western colonialism in Asia (Tu, 2000; Chen, 2018), they now have drastically different economic and political systems. These systems lead to heterogeneous cultural practices and media use, as well as different industry structures and policies. Second, even

4

though this region gives rise to some quintessentially 'Asian' media technologies such as the pachinko game (see Chapter 9), esports (see Chapters 7 and 8), and simulated images of nuclear weapon testing (see Chapter 6), the three countries encounter global issues experienced by all digital societies, from data privacy to media piracy, from the commodification of online activities to the commercialisation of culture. Third, local and global forces both shape the market and cultural practices of media technologies. Even citizens in North Korea, one of the most secluded countries in the world, cannot avoid alluring global images and technologies such as Hollywood films and smartphone technology. What this means is that media technologies for work and play in Northeast Asia are always shaped by local and global forces in a historical context.

To further explore the tensions between work and play, and between Asia and the rest of the world, and to further illustrate the aforementioned three points, this edited volume aims to: First, discuss how a political–economic approach informs cultural and technological approaches and vice versa in the study of media technologies for work and play in Northeast Asia; and second, discuss how the peculiar geopolitical relations between Japan, South Korea, and North Korea can be understood from the standpoint of media technologies for work and play. The following sections will detail these two aims.

First aim of the book: Multiple approaches to study media technologies

How can we study media technologies that blur the boundary between work and play? Work has traditionally been more valued than play in Japanese and Korean cultures (Bossy, 2000; Hu, 2015). Elders and parents have long seen media technologies for play as harmful to youngsters as well as society at large (see Chapter 4). Play is seen as a frivolous pursuit that gets into the way of serious work. Children are only allowed to watch television or play video games after they finish homework and household chores. However, when technologies for play become a viable part of the digital economy, game development is seen as a proper profession for young people. Moreover, when governments and corporations invest in media technologies such as esports, playing video games is more than an inconsequential trivial pursuit, but is a political–economic issue for politicians, investors, and designers. Lawmakers want to debate how the industry and video gaming can be regulated; investors want to make money from a growing industry; and content producers want to create more and better products. Because of the various stakeholders with different interests, media technologies for play are no longer just about culture, but also about politics, economies, and technologies. How can we best understand such a complicated phenomenon?

We argue throughout this book that no single perspective can grapple with the complexity of media technologies for work and play in Northeast Asia. We will illustrate how the three perspectives – political-economic, cultural, and technological – enable a comprehensive examination of media technologies for work and play. In the following, we will again use esports as an example, because this was born from and relies on a particular political-economic environment, shared culture, and advanced digital technologies. Therefore, multiple lenses are needed to explain its popularity in Asia. While a political-economic approach is effective at looking at the industry structure and labour conditions, a cultural approach is effective at understanding how local populations interpret the content, and a technological approach is effective at understanding how technologies are interpreted. In the following, we explore the questions asked from each of the three approaches.

A political-economic lens sees esports as an industry in which pro gamers are workers who are paid to produce entertainment for the audience (Jin, 2010). Entertainment is seen as a useful social good because it provides a leisure pastime to viewers. Gamers and esports spectators are willing to pay for such goods. Seeing the demand for esports, investors put their money in it hoping to make a profit. Investors can make money from owning teams, developing popular games, and enhancing brand images through advertising and sponsorship. For example, South Korean professional esports teams are sponsored by conglomerates in the telecommunication, entertainment and aviation industries. Sponsorship is the largest funding source in the global esports market, accounting for more than 50 per cent of revenue. In return, sponsors such as SK Telecom, Samsung, CJ Entertainment, and Korean Telecommunication can enjoy branding effects, making their names known not only in the country, but also worldwide.

Professional gamers may appear to be heroes to fans, but they are workers who sell their labour to team owners. Professional players go through a competitive selection process. They first start as gamers competing for major esports titles. Their 'work' in playing *League of Legends*, *Starcraft*, or other esports titles is only recognised when they emerge as top players in matches hosted by esports game companies, Korean government offices, or the owners of PC *bang* (Internet café). The paid value of professional players' work represents an intensive training schedule to further improve team playing skills under the close monitoring of coaches and corporate sponsors. While fans may pay more attention to the superstars who are top earners, they pay less attention to the income disparity between the top earners and average workers. In 2019, the top esports player made 43 times more money than the one ranked 50th; the former made US$1.2 million and the latter made less than US$30,000 in a year (Esports Earnings, 2019). In other words, the top player earns as much as the CEO of a public company while the player ranked 50th only makes as much as someone working in a coffee shop. In this sense, the inequality of labour conditions in esports is similar to many industries such as professional sports and entertainment.

Another similarity to professional sports and the entertainment industries is that problems such as gender inequality and job scarcity persist. There tend to be more male players than female. Women are also paid less because they are believed to draw fewer viewers. In addition, they tend to be more sexualised than men. Female players in the South Korean esports team IM Athena epitomise the appearance of women in esports: The team image is feminine and innocent, framing the women as part of the *kawaii* (cute) culture in computer gaming. There is also the problem of job scarcity; even though esports is a fast-growing industry, stable jobs are hard to come by. Many professional players have to retire in their mid-twenties because their muscle strength and reflex start to deteriorate, making it harder for them to compete in speed.

Again like the sports and media industries, pro gaming is not a standalone industry; its success depends on other industries, such as video game development, telecommunications, and mobile technologies, as well as on advertising and marketing, merchandising, media, and venue construction and design. Besides, the success of the pro-gaming industry depends on the fans who discuss the games in person and online, as well as on the enthusiasts who create videos and write blog posts. When online activities can be easily tracked, fans provide digital labour (Fuchs, 2012; Scholz, 2013) that can be potentially commodified by marketers and big data firms. In other words, while fans are watching the games online, what they watch and how they watch it provide data for platform and telecommunications companies to learn about users' behaviours (Boutilier, 2015). These data can then be repackaged as consumer analysis and sold to advertisers and marketers to help them better target consumers.

But a political-economic viewpoint does not take the cultural aspect of esports into consideration. From a cultural perspective, esports challenges cultural values ingrained in Japanese and Korean cultures. Esports does not only challenge these values, it also reflects the tension between work and play in the broader context of the digital economy. Filial piety guides human relations in school and the workplace in Japan and the Koreas. The Asian workplace is hierarchical: Subordinates are expected to submit to superiors who will in turn assume responsibility for subordinates (Lincoln and Kalleberg, 1985). This hierarchical relationship is unlike that in the 'flat', egalitarian workplace of the digital economy, where subordinates are encouraged to challenge superiors (Craig, 2018). In addition, superiors in a flat workplace do not assume sole responsibility because everyone in the team is expected to contribute to projects. Will the work culture of the digital economy challenge Asia's hierarchical workplaces? Or will workplace hierarchy constrain workers' creativity in the digital economy, hindering countries from fully embracing media technology development? On a related note, if the digital economy can change workplace culture, can it also change cultures in schools and households, permitting students to challenge teachers and children to challenge parents? To satisfactorily answer these questions, it is important to learn how people in these three societies interpret social norms when they use media technology in digital gaming.

The digital economy also gives a new meaning to employment in the three countries, creating a work–play tension between the promise of lifetime employment in Japanese and Korean companies and precarious employment in the digital economy. Traditionally, Japanese and South Korean workers have been guaranteed lifetime employment and have been expected to completely devote their waking hours to the company. To show loyalty to the employers, many workers regularly spend more than eighty hours at work, at the cost of their health. The Japanese word 'karōshi' is used to describe death resulting from overwork. In 2018, South Korea passed a law prohibiting employees from working more than fifty-two hours a week. In North Korea, state workers are expected to show up to worksites even if there is no work to do and wages are not sufficient to meet basic needs (Hastings, 2016).

However, the bursting of the Japanese economic bubble in 1980s and the Korean economic crisis in 1990s have restructured work organisations, making lifetime employment a thing of the past. Few workers these days are expected to stay in one company for life, and some are not even expected to stay in the same industry. The digital economy that proliferated after the burst of the bubble favours freelance, flexible, and temporary jobs. Workers, on the surface, have more freedom to try out different kinds of jobs and industries.

However, these flexible jobs are mostly dead-end, low-paying jobs, and do not provide the lifelong security that a traditional workplace does. While Japanese and Korean women were disadvantaged in traditional workplaces, they also did not gain much in the flexible economy (see Chapter 1 and introduction to Part I).

The digital economy may also give education a new meaning, creating a tension between playing and studying. Japanese and South Korean school children are expected to study long and hard for college entrance examinations (Bossy, 2000; Hu, 2015). Students commonly attend cram schools after regular schools. Adults see play technologies as a distraction, so children can only enjoy them when they are done with serious work. However, adults also use media technologies to reward good academic performance. Adults then see play technologies as both a temptation and a bribe; they need to be restrained and manipulated at the same time. 'Studying' esports as a school subject therefore seems to provide a happy compromise between parents and children. Since esports is now a legitimate profession in South Korea and an area of study in university (Denyer and Kim, 2019), many young people see it as a viable option in case they do not get into prestigious fields such as medicine or law.

The economic benefits brought by media technology development also encourage governments and corporations in Japan and South Korea to invest heavily in the ICT and video game industries, especially if these technologies can also be exported overseas to boost domestic economies. Two good examples are Japan's Nintendo and South Korea's Samsung. The products of both brands are considered culturally odourless (Iwabuchi, 2002), they are not *too* Japanese or Korean and they do not confuse overseas users. The sales figures illustrate the transnational appeal of these brands. From March 2018 to March 2019, 76 per cent of all Nintendo products sales came from the overseas market (Nintendo, n.d.). Similarly, 90 per cent of Samsung's revenue came from the international market (Asif, 2017).

To conclude, a cultural lens brings to the forefront the tension between societal expectation of work and place. This tension is brought by economic downturns in Japan and South Korea. To cope with youth unemployment, governments and corporations promote a digital economy, believing it to be a solution to a weak labour market (McRobbie, 2016). However, the digital economy may exacerbate the problem because its design prioritises state and corporate interests.

Adding to the political-economic and cultural perspectives is a technological one. Media studies literature that is informed by science and technology studies (Gillespie et al, 2014) argues that technical devices are not simply conduits of content; users interpret the devices as much as they interpret the media content enabled by them. In game studies, technologies could enable certain content that adds value for play or constrain others (Flanagan and Nissenbaum, 2014). Using the esports example once again, we argue that users interpret a video game played on a desktop computer differently from the same game played on a handheld console. For example, a fast Internet connection is required for competitive esports, so an esports game played on a console may not be as exciting as it is on a desktop, even if the content is the same. On the other hand, Nintendo designs some games that are *only* appropriate for handheld consoles. For example, some game content requires the devices to have the motion-sensing feature that desktop computers lack. This may explain why *Pokémon GO* was not designed for PC. In addition, handheld console games are supposed to be played 'on the go', so the game design allows for short, casual play that is used to fill time while waiting for real life to move on (see Chapter 9 for how pachinko was a game people used to play to kill time).

The technological perspective also argues that technologies are not politically, economically, and culturally neutral. The development and design of technologies are already embedded with values (Wajcman, 1991). The case of North Korea, though extreme, is a good illustration of the interdependence between content and technology. The North Korean government sees the media as a propaganda tool, so it owns, governs, and regulates all mass media technologies and content. Mass communication is designed for public consumption, not private consumption at home. For example, newspapers are displayed in public places (such as train stations) so that citizens can read them in public (*The Telegraph*, 2018). Mobile phones and email accounts are not widely owned, not only because of the financial hardship of the average North Korean, but also because these technologies enable one-on-one communication, which is seen a threat the power of the regime (Martin and Chomchuen, 2017). Therefore, communication on mobile phones and emails are closely monitored by the government. Another example of the interdependence between technologies and content in North Korea is the limited number of websites registered with the North Korean domain site '.kp' (Burgess, 2016). The limited number of websites has little to do with cost because webpages are much cheaper to produce than print newspapers or films. It also has little to do with the lack of technological know-how, because there are knowledgeable software engineers in the country (Jun et al, 2015). The few websites registered as '.kp' may have a lot to do with the dearth of suitable content that *requires* the online platform, because North Korean citizens are not allowed to access the Internet, let alone being able to produce online content and upload it.

As we will show in the literature review, current studies on media technologies in Japan and the Koreas tend to focus on either the technologies or content, but not on how content is interpreted because of the technologies that enable it. Current literature on media technologies in Japan and South Korea also reflect the silos between academic disciplines: Scholars in business and public policy tend to focus on the broader political-economic environment by attending to media technologies; scholars in cultural studies tend to focus on content such as anime, manga and video games, and less on the affordances of technologies. To remedy the divide, chapters in this edited volume look into the hybrid and cross-border mediated space where East Asian culture and identity shape its various forms in the three societies.

Second aim of the book: Geopolitical relations in Northeast Asia

The second aim of this volume is to discuss how media technologies in Asia illustrate the peculiar geopolitical relations between Japan, South Korea, and North Korea. Geopolitics is preoccupied with borders, resources, flows, territories, and identities. It offers analysts a way to understand the global landscape using descriptions and metaphors (Dodds, 2007).

Current literature on media technologies tends to see Asia as a unique region. Moreover, scholars on Asian regionalism tend to focus on one single Asian nation in their analyses. We believe contemporary Asia needs to be understood in the geopolitical relationship in Northeast Asia that is shaped by Cold War politics (Otmazgin, 2013). This relationship in turn has implications for the different stages of technological development and consumption practices in the three countries. The inter-regional media network among the three countries is shaped by the availability (or lack) of raw materials and human power, which in turn reflects how the three nations see themselves and each other as they cope with their unique historical pasts since the beginning of the twentieth century, from colonisation to the two world wars, from the Cold War to globalisation, from economic prosperity to recession. A focus on geopolitical relationships will help avoid the pitfall of seeing 'the commodity-image-sound of mass culture becomes the fundamental form in which the putative unity of Asia is imagined and regulated' (Ching, 2000: 235). By no means is this the only geopolitical dynamic shaped by the Cold War; Western Europe and Eastern Europe is another example. However, Northeast Asia is unique in the sense that it has two of the wealthiest and technologically most advanced countries in the world, alongside one of the poorest and the least advanced. It also has two countries – North and South Korea – that are officially still at war with each other.

In most global media studies literature, geopolitical relationships are nation-centric. What this means is that this relationship is one between nation-states rather than between cultures (Park and Curran, 2000). A nation-centric perspective raises an interesting question on the way to defining Asia from a single national point of view (Ge, 2010). J.B. Choi (2010) believes that the boundary of nation-state does not stop the construction of geocultural spheres in the production and circulation of culture. Geopolitical relations may facilitate political-economic and cultural collaboration, but may also create conflicts. An understanding of this geopolitical relationship necessitates situating media technologies in a broader historical background that illustrates past and ongoing push and pull factors such as colonialism and independence, war and reconstruction, Westernisation and indigenisation, and state-led development and globalisation. These four pairs of factors will be explained later in this chapter.

Returning to the case of esports, peculiar geopolitical relations in Asia are reflected in the countries that sent representatives to compete in esports and those that did not. The only country from Asia that did not send representatives to esports was North Korea. The reason is not surprising: Broadband Internet connections are not available to the majority of the population due to the regime's unwillingness to risk losing political control over its populations. This complete absence of broadband connections in North Korea sharply contrasts with Japan and South Korea, where the populations enjoy fast and cheap Internet access. In 2018, Japan and South Korea ranked 12th and 30th respectively in the world for fastest fixed broadband access (Lai, 2018). To understand why North Korean political leaders have a tight grip on citizens' access to information while Japanese and South Korean leaders promote digital technologies, it is important to understand the trajectory of how the three countries modernised in the twentieth century, which will be laid out in the next section.

Complicating the post-Cold War geopolitical relations of the three countries are the political, economic, and cultural influences from the US and China. In different historical eras, the US and China brought political structure, cultural influences, and commercial interests to

Northeast Asia. Prior to the twentieth century, court officials and merchants from different Chinese dynasties brought religions, languages, and trade to the region. However, since the conclusion of World War II, the US has been influencing Northeast Asia in a more pronounced way because of the power struggle between the US and the USSR.

The defeat of Japan and Germany in World War II paved the way for the Cold War that subsequently divided the world into two blocs: the Western allies led by the US and the communist bloc led by the Soviets. The geopolitical relations in Northeast Asia were redefined during the Cold War: The US reconstructed post-war Japan and aided South Korean economic development, while the Soviet Union and China aided North Korea after liberalisation. In addition, the US set up multiple military bases in South Korea and Japan, ready to strike if North Korea invades its southern and eastern neighbours.

In the late 1980s, the geopolitical relations among the three countries were once again redefined when the Soviet Union collapsed and withdrew financial and political support for North Korea. The withdrawal of support significantly slowed down North Korea's economic development, along with telecommunications development (Clark, 2012). China, another ally of North Korea, pursued a dramatically different economic development path at the end of the Cold War. It implemented the Open Door Policy and reintegrated its economy with the world economy. The collapse of USSR and the opening of China made North Korea become one of the most isolated communist countries in the world.

Even though the Japanese colonisation of the Korean peninsula has made Koreans suspicious of Japanese imports, the global economy and trade liberalisation have soothed some of the historic anger. For example, the uneven economic development between Japan and South Korea has promoted economic collaboration. Japanese animation companies took advantage of cheaper labour costs in South Korea and hired offshore artists to do their sketching (Kim, 2014). South Korea also has taken advantage of its cheaper labour to produce quality electronic goods for the global market. Japanese electronic brands (such as Sony, Panasonic, Sanyo, and Sharp) once dominated the overseas markets for television sets, radios, and camcorders, but they lost their competitive edge when South Korean firms such as LG began to produce cheaper electronic goods for the global market.

The governments of Japan and the Koreas also see economic collaboration as a strategy to ease political tensions. In 1998, the late President of South Korea, Kim Dae Jung implemented the Sunshine

Policy to encourage small-scale collaboration between South Korean firms and the North Korean government (Kim, 2004). For example, the Kaesong Industrial Region in North Korea was set up to facilitate technology transfer from the South to the North. This region that borders South Korea was an experiment to test the possibility of peaceful co-existence of the two countries.

There are more examples to show how post-Cold War geopolitics has shaped media technologies in Northeast Asia. Without a doubt, the current conflicts between China and the US will shape the direction of geopolitics in the region. By the time of writing, both Japan and South Korea appear to reflect complex Northeast Asian geopolitics, with South Korea maintaining its connection to China in order to control North Korea, while Japan chooses to support the US in the hope of regaining political influence in the Asia Pacific region.

In the following section, we will step back and delve further into history by examining four pull and push factors that have shaped the geopolitical relations in this region.

Four pull and push forces: Japan and Korea in the long twentieth century

A historical perspective is required to advance the two aims of this book and to avoid the pitfalls of seeing Asian technologies for work and play as exceptions to Western practice. To reiterate, this edited book aims to use multiple approaches to examine technologies in Asia: While Japan and the Koreas share some common cultural features, such as Buddhism, political-economic tensions in the twentieth century have forced these three countries to respond to sweeping changes through and with media technologies in different ways. The tensions with traditional culture and sweeping changes in turn shaped and reshaped the geopolitical tensions in Northeast Asia. Media technologies sometimes intensify the tensions and sometimes soothe historical scars. By situating the development and use of media technologies for work and play in Japan and the Koreas in a historical perspective, we avoid seeing technological development as a linear process dominated by the West. We refuse to see the Western way of using technologies as the norm to which Asians must conform. Even though Asian countries have historically responded to technologies that originated in the West, these countries have also shaped technologies for nation-building and economic development. As such, technologies in East Asia cannot be compared to those in the West without situating these comparisons using a historical perspective.

The four pairs of pull and push forces are: (1) colonisation and independence; (2) war and reconstruction; (3) Westernisation and indigenisation; and (4) state-led development and globalisation. These forces have shaped the development and use of media technologies for work and play, which in turn have contributed to the economic, political, and cultural development of the region in the twentieth century.

Colonialism and independence

The Korean peninsula is mountainous and resource-poor, and has historically looked to China to protect it from foreign invasion. In the nineteenth century, the weakening of China prompted Japan to assume a more prominent political-economic role in the region. In the late nineteenth century, the Japanese forced the Koreans to sign a treaty to open trade and diplomatic relations. The subsequent defeat of China in the Sino-Japanese War forced China to retreat from Korea (Clark, 2012). From 1910 to the end of World War II, the Korean peninsula was annexed to Japan as a colony.

Korea's coloniser Japan stood out among all Asian countries not only because it had never been a colony, but because it had become a coloniser of other Asian countries. Imperial Japan country occupied Northeast China, Taiwan, and Korea seeking raw materials and imposing trade advantages. While the economic development of the Japanese colonies primarily concentrated on agricultural production, Japan also engaged in empire-building by constructing large-scale industrial infrastructure. Raw resources such as metals extracted from colonies bolstered Japan's self-sufficiency for war preparation (Uchida, 2012). Trade and colonisation required modern communication technologies, leading Japan to build communications infrastructure such as telephone and broadcasting systems in the Korean Peninsula. In addition, the Japanese government believed that radio broadcasting was an effective means to eliminate local Korean culture and promote Japanese culture in the peninsula.

The Korean Peninsula gained independence after the Japanese Emperor surrendered at the end of World War II in 1945. However, Korea did become not immediately politically and economically independent. After the outbreak of the Korean War (which will be covered in more details in the next section), the peninsula was divided into two countries: The Republic of Korea (thereafter South Korea) was established on 15 August 1948 and the Democratic People's Republic of Korea (thereafter North Korea) was established less than a month later. Both newly founded countries were politically and economically dependent on the Cold War factions that they allied with: South Korea formed an alliance with the US, while North Korea sided with the USSR and the People's Republic of China (PRC).

The fates of North and South Korea soon diverged due to national ideologies and the collapse of the USSR in 1989. Media technologies played a prominent role in South Korea's quest for economic prosperity and political independence. In the 1980s, the South Korean government was determined to transform its economy from exporting low-value goods (such as household items, garments, and low-end electronic goods) to high-value goods (such as audiovisual technology and computers). Besides, it saw telecommunications infrastructure as a means of economic development (see Foreword 2). The investment paid off. Within a short span of three decades, South Korea lifted itself up from being a developing country to one of the most economically developed and digitally connected countries in Asia.

In contrast, the Korean War and the withdrawal of Soviet economic support left North Korea's communications systems in shambles. The country did not invest heavily in communications systems because the Korean Wave left the country with few allies and the collapse of the USSR further isolated it. While present-day North Korea is a politically independent country, its weak economic position left it with few choices in terms of media technology development. Currently it has to rely on 'non-hostile' countries to import technologies, such as cellphone networks from Egypt or 'notels' (notebook computers and televisions) from China.

War and reconstruction

World War II redrew power struggles in Northeast Asia. Although Japan went head to head with the Allies during the war, afterwards it became a close US ally. The development of media technologies in Japan can be contextualised in terms of both Japan's determination to rebuild the country and its admiration of Western technologies.

In 1945, atomic bombs were dropped on Hiroshima and Nagasaki, marking the end of World War II in Asia. After the Emperor surrendered, the US military occupied Japan and brought democracy to the country, helping it to write a new constitution that stipulated working conditions, reformed education, and brought suffrage to women. Economically, Japan had to rebuild its industry after losing a quarter of its national wealth during the war. The Japanese government used media technologies to bolster the national economy by repurposing military facilities to produce electronic goods, such as cameras, for export (Pile, 1996).

After being liberated from Japan, South Korea ignited nationalism in its population to speed up economic recovery through media technologies. The South Korean government vowed not to import or adopt Japanese technologies, and instead looked to the US for technology transfer and training (Kim, 1996). The government also sent students to the US to learn the latest technologies.

One outcome of the Korean government's resistance to Japanese influence was the different trajectories of media technologies development in South Korea and Japan. For example, the South Koreans pioneered massive online gaming. while the Japanese pioneered handheld game consoles. The South Korean firm Samsung develops products with the global market in mind, while Japanese cellphone firms mainly target domestic markets.

Westernisation and indigenisation

Contemporary Japan and South Korea are seen as modern Asian countries. However, their paths to modernisation and Westernisation have been different. North Korea is often considered not to be modern at all. Among the three countries, Japan was the first to modernise itself after a long closed-door period prior to the Meiji era. Militarisation during the Meiji era also fuelled Japanese ambitions to colonise the Korean peninsula, during which the colonisers brought modernisation to the colonised. After the outbreak of the Korean War, North Korea adopted extreme policies to reject Westernisation.

The Meiji Revolution opened the doors of Japan and the Emperor looked to the US and Europe to modernise and Westernise the country. To achieve the goal of nation-building, the Japanese government sent students abroad to Europe, and these educated men returned to Japan to lead the modernisation effort (Ahmed, 1988). The Japanese government also actively brought in foreign experts to work in the Department of Industry, which was in charge of transport and communications. The majority of foreign nationals who worked in telegraphs and telephones came from England and France (Shirahata, 1998).

By comparison, Korean technological development in telegrams and telephony in the early twentieth century was not a result of the country's desire to learn from the West, but was rather imposed on it by its coloniser, Japan. After the Korean War, an impoverished South Korea saw that its communication technologies, such as telephony and television, lagged behind Japan both in terms of penetration and international standards. Even in the 1980s, when its economy took off, landline telephone ownership in South Korea was lower than television ownership, because telephone ownership was strictly restricted by the government (Larson, 1995; Oh and Larson, 2011).

Across the Sea of Japan, North Korean leaders were rejecting Japanese and Western cultures altogether. The state adopted a self-reliance (*Juche*) ideology that condemns economic, political, and cultural influence from the West and Japan. This ideology bans the distribution and consumption of foreign media in the country. Citizens who are found to smuggle in, possess, or consume foreign media could be sent to labour camps, if not executed (Freedom House, 2013).

The opposite of Westernisation is indigenisation. Indigenisation refers to the process of creating culture and technologies that fit local tastes, intended for local use (Cleveland and Bartsch, 2018). Although indigenisation may be seen as a process of excluding foreign culture, cultural producers can draw on foreign materials and mould them to fit local contexts. For example, even though the Japanese learnt filmmaking techniques and imported equipment from the West, world-renowned Japanese film directors Akira Kurosawa (1910–1998) and Yasujirō Ozu (1903–1963) developed cinematic styles that did not imitate Hollywood aesthetics. Ozu invented the 'tatami shot' by lowering the camera in a way that suitably frames characters sitting in *seiza* by the table. Another example of indigenisation is Hayao Miyazaki's two-dimensional animation. Although Miyazaki was inspired by Walt Disney's earliest films, he did not change the aesthetics of his animation when Disney developed three-dimensional animation aesthetics.

North Korea is an extreme case of media technology indigenisation because the leaders deem Western culture (including South Korean culture) to be pollutants that will harm the purity of the Korean race. Even though cellphones and the Internet are available to a very limited portion of the population, their way of using these technologies is completely opposite to the West's. For example, in the West the Internet's inception was as a decentralised network that could escape the control of governments and corporations, but the North Korean government strictly controls who can access the Internet and what communications can be exchanged. Instead of joining the borderless Internet, the government instead developed an intranet which, like other state-owned media, only distributes state-approved content.

State-led development and globalisation

The last tension is between state-led development and globalisation of media technologies. Globalisation refers to the accelerated flows of money, ideas, technologies, and people around the world (Eitzen and Zinn, 2012). The actors that initiate the flow can be international organisations, national governments, and transnational corporations. Media and information companies from Disney to Google have the power to homogenise media products and services because they tend to sell the same goods in different markets. Ordinary citizens can also facilitate the flows through migration, travel, and business. International and intraregional organisations (such as the European Union and the United Nations) can also initiate the flows through global summits.

One consequence of the accelerating flows of goods, ideas, and people is the spread of neoliberalism, an ideology that suggests that local governments are no longer relevant to the everyday lives of the populations because private entities such as transnational corporations can provide their needs and wants, from healthcare and education to job training and transportation (Harvey, 2007). This ideology encourages transnational corporations to take on some of the roles that local governments have traditionally played. For example, the government is traditionally seen to be the body that encourages technological invention, but transnational corporations can negotiate with local governments to discourage local invention and to encourage imports of foreign technologies.

Although globalisation is seen as the opposite of state-led development, the cases of Japan and South Korea show that both forces can co-exist. In this sense, technological development policies are unlike those in the US and Canada, where the governments do not have much of a role in planning large-scale media technology development. The governments of Japan and South Korea not only plan what media technology sectors to develop, but they also invest in infrastructure and private-public partnerships. In 2012, Japan launched the Cool Japan project to build the country's soft power by promoting *kawaii* culture abroad (Ministry of Economy, Trade, and Industry, 2012). In South Korea, the government played a major role in successfully transforming a developing country into a digital society (Jin, 2017; see also Foreword 2). The government handpicked Samsung and LG, two Korean *chaebols* (family-owned conglomerates) to develop media technologies such as television sets and smartphones for both the domestic and international markets. But the government also limits how free the competition is by making sure that *chaebols* do not directly compete with each other in all industries (Oh and Larson, 2011). This level of regulation is unimaginable in countries that adopt neoliberal policies, such as the US and Canada.

However, state-led development of media technologies does not preclude foreign brands from influencing local cultures. For example, South Korea had actively resisted the import of Apple's iPhone, but once it was introduced, it drastically changed South Korean consumers' expectations of smartphones (Lee, 2012). Similarly, even though North Korean leaders overtly resist globalisation and punish citizens for consuming global media goods, the government could not stop citizens from actively seeking foreign media goods and technologies on the black market (Zhang and Lee, 2019; see also Chapter 5). If one of the world's most isolated nations could not rein in global media technology forces, then Northeast Asia is hardly an exception to a global culture.

After reviewing the four pairs of push-and-pull forces that create tensions in the development and use of media technologies in Northeast Asia, we reiterate that no one single approach (political-economic, cultural, or technological) can adequately examine media technologies for work and play in Japan and the two Koreas. In addition, this book recognises that geopolitical relations in Northeast Asia condition how technologies are developed, used, and understood in the region. This book further recognises that the practices of technology are locally specific, shaped by historical circumstances, political-economic arrangements, global influences, and cultural understanding. The book also recognises that the rapidly developing social, cultural, and economic realms in Northeast Asia constitute a new geopolitical relationship in the Asia Pacific Rim. The complexity in media development not only speaks about the region's past since the Cold War era, but it also sheds light on the potential for a New Cold War, because the link between Japan and the Koreas points to a new political mapping between the world's two economic superpowers, the US and China.

Before we review previous literature on media technologies in Japan and the Koreas, we need to state what this book is *not* about: First, it does not treat technology use in the three East Asian countries as an exception to Western countries; second, it does not see the three countries as a comparative case to the West. In other words, the West is not seen as the norm from which Asian countries deviate and with which Asian countries need to catch up. By not seeing Asia as an exception or comparison to the West, this book avoids attracting the Western gaze at Asian technologies. What this means is that some media technologies examined in this book (such as adult PC games and pachinko in Japan) are not treated as examples of peculiar cultural differences between the East and the West, or as instances that suggest why the East is not the West. Although these technologies may not be popular outside these countries, they are meant to show how global currents such as consumer culture, the feminisation of the economy, and precarious digital labour manifest themselves in different forms. For example, surplus labour in the digital economy is absorbed in different ways in South Korea and the US: When massive online video gaming does the trick in South Korea, media internships in the US manage surplus labour (Corrigan, 2015). By not seeing Asia as an exception or a comparison, this book acknowledges that technological development is neither monolithic nor unidirectional; it is always tied to the history, culture, geography, and political economy of the countries and regions. As such, technological developments in different countries are mutually constituted by how the populations make sense of their material practices.

Current literature on media technologies in Japan and the Koreas

This section reviews the literature on media technologies for work and play in Japan and the Koreas. The literature can be categorised into two groups. The first examines technological development in Japan and South Korea since the 1970s, asking why East Asia has caught up with – if not surpassed – Western countries in such a short time. The second group studies how Koreans and Japanese use technologies in a local context that reflects new digital cultures. There is little conversation between scholars in these two groups. The linkage between a macro context and micro, local practices is often unexplored.

When compared to the literature on South Korean and Japanese technologies, there is a dearth of studies on North Korean media technologies. The few studies that touch upon media technologies in North Korea also have a different focus from those on Japan and South Korea; they tend to seek an understanding of how the government controls citizens' access to technologies and, in response, how citizens turn to the black market for underground technologies. As a whole, current studies do not explore the technological dynamics in Northeast Asia.

The first group of studies explained why the two countries rapidly developed the Internet, mobile telephony infrastructure, and consumer products since the 1950s. Scholars commonly agreed that the path of development in Asia is different from that in Western countries because of political will and cultural background. In contrast to Western countries, the governments of Japan and South Korea played a significant role in planning initiatives and implementing regulations (Sternberg, 2003; Hemmert, 2007; Oh and Larson, 2011; Holroyd and Coates, 2012). Scholars also believe that East Asian culture and geography provide favourable conditions for rapid technological development, such as an emphasis on education (Larson, 1995), demographic composition, and a concentrated population in urban areas (J.H-j. Choi, 2010).

The examination of speedy technological development inevitably compares the telecommunications and computing sectors in South Korea and Japan with those in Western countries, in particular the US. Scholars compare areas such as legislative changes (Larson, 1995; Nishioka and Sugaya, 2014; Yoo and Lee, 2014) and their impact on telecommunications liberalisation. They also compared telecommunications services in Japan, South Korea, and the West in terms of costs, connectivity, and speed. Scholars commonly agreed that Asian telecommunications services are cheaper, faster, and have better coverage (Larson, 1995; Holroyd and Coates, 2012). The speedy development of telecommunications is said to have transformed Asian countries in a short time (on South Korea, see Oh and Larson, 2011), economically integrated the region, and challenged Western stereotypes about Asians as being technically backward and having little design aptitude (Holroyd and Coates, 2012).

The second group of literature looks at the technological culture for work and play in South Korea and Japan. Scholars agree that technologies have to be understood in specific sociocultural contexts because cultural and social filters affect how technology is represented, imagined, and appropriated (Yoon, 2010). Chan (2013) adds that nation-specific cultural nuances differentiate the production, circulation, and consumption of new digital media in the region and elsewhere. In the case of post-World War II Japan, the economic success of industries such as household electronic appliances, computing, anime, and so on led to the development of the video game industry in the 1970s (Picard, 2013). In other examples, US-originated media technologies such as the iPhone (Lee, 2012) and YouTube (Kim, 2013) have impacted local cultures. Lee (2012) suggested that the iPhone remediated the use of existing media such as the mobile phone, Internet networking, and multimedia play among young people. While most studies focus on media technologies and telecommunications since the 1970s, a few have also paid attention to historical cases such as how the Japanese used radio broadcasting to control the Koreans during colonisation (Kim, 2015).

Some studies look at how technologies for work and play reinforce traditional Asian cultures while creating new ones. For example, in contemporary Japan, mobile phones enhance family lives by bringing children closer to their parents, but they have also created a new sense of vulnerability for children (Spry, 2010). Mobile phones have also created a new culture in Japan for young people, such as creating a new female identity (Hjorth, 2010), and new slangs and symbols (Matsuda, 2010). In South Korea, mobile phones have created a new population called transyouths who connect to each other through screens and PC *bangs* (J.H-j. Choi, 2010).

Current literature often implicitly compares Asian cultural practices of technologies for work and play with those of their Western counterparts. For example, South Korean youths were found to have smaller social circles but stronger ties with people in their social networks in comparison to Western youths (Gallagher, 2017). Japanese video game arts and anime were also said to not follow Western aesthetics. Japanese animators prefer two-dimensional characters and juxtapose them on a three-dimensional landscape in order to counter an assumed 'natural' media technological advancement (Saito, 2013). Another study compared how election campaigns in Japan, South Korea, and US used the Internet (Kiyohara et al, 2018). It found that Asian political campaigns are less effective at using online platforms than campaigns in the US, which is seen as a model of democracy and civic engagement.

Studies on media technologies in North Korea have a very different focus from those examining Japan and South Korea. Some focused on how the government uses technologies to survey and control the populations, others focused on how citizens use homemade and black-market technologies to resist the power exerted by the government. Jun, LaFoy, and Sohn (2015) studied North Korean cyber operations and named the government bureaus that may be responsible for them; Kim (2004), Ko, Lee, and Jang (2009), and Ziccardi (2013) looked at the highly restrictive Internet technology in North Korea; Noland (2008) looked at telecommunications development in North Korea. Scholars that looked at technologies obtained on the black market include Baek (2016), who conducted in-depth interviews with North Koreans who reside in South Korea, and Larson (1995) who speculated as to how a unified Korea might impact telecommunications development in the two Koreas. Larson suggested that unification will create a larger consumer and labour markets, and will connect North Korea to Russia, Western China, and Mongolia. Since few researchers have direct knowledge of media technologies in North Korea, all the studies on the country tend to be tentative.

This edited volume aims to enrich current literature by providing a holistic understanding of media technologies for work and play in Northeast Asia by connecting a broad political-economic framework to local cultural practices. In addition, it seeks to understand the dynamics among the three countries by looking at how technological development and practices are informed by specific historical political-economic contexts.

Overview of the book's chapters

This volume brings together ten original chapters on media technologies for work and play in Japan and the two Koreas. We organise the chapters into three parts: 'Gender Online and Digital Sex', 'Governance and Regulations', and 'Digital Labour'. A mini-introduction opens each part, offering readers background information about some common issues and connections among the essays. The mini-introductions are recommended for readers who have little background in a critical understanding of media technologies in Japan and the Koreas. Instructors who assign course readings can also choose to assign the mini-introductions and associated chapters if this introduction proves too expansive.

We must note that some of the chapters fit into more than one part. Also, we note that even though some parts may have chapters on the three countries, by no means are they comparable cases. As previously stated, we do not believe that technological development is linear; how technologies are produced, regulated, and used very often manifest localised issues such as audience subjectivity and meaning-making, labour arrangements, and industry structure.

Part I begins with Kyooeun Jang's examination of how South Korean underemployed youths cope with a stagnating labour market. In Chapter 1, she looks at how young women were encouraged to become financially independent by becoming online entrepreneurs. Jang showed that these young women very often explain their business successes with their passion and their failures with their own lack of grit. Jang argued that instead of being free and independent, online entrepreneurs have to sell affective labour as well as commercial goods online. They also sexualise themselves regardless of the goods they sell. For example, owners of floral arrangement shops wear bathing suits to sell their products. Lastly, online businesses are seen as a way for mothers to rejoin the workforce while remaining dutiful mothers and housewives. A number of mother-entrepreneurs frequently post pictures on social media sites of themselves working while taking care of young children.

Patrick Galbraith examines adult video games in Chapter 2. Non-Japanese people often regard adult video games in Japan as perverted media that is evidence of how weird and odd all Japanese people are. Galbraith first argues that adult video games are actually a niche genre. The video game producers and consumers form a close-knit community that is suspicious of non-Japanese out of the fear that Western media will widely report such games. In the past, Western reports have caused the international community to put pressure on the Japanese government to stamp out the genre. Some believe that it would be better if the games were played only by Japanese people inside the country. However, some game developers and consumers are keen to expand the community outside Japan. To understand these two opposing directions, Galbraith questions the extent to which a game can be contained within a country. Even though the games are deliberately designed for Japanese-speaking customers living in Japan, they spill over the border thanks to participants who believe that the community can only thrive when the games are more globalised.

Part II is about governance and regulations. Laws and regulations are written to govern local populations; local cultures and histories in turn shape how they are written. However, globalisation means national governments have to conform to international standards at the expense of forcing local populations to become global subjects. In Chapter 3, Ana Gascón Marcén's explores how Japan and South Korea responded to the demands from the European Union to strengthen laws on data protection. Marcén argues that Japan lags behind European standards because during World War II, organised neighbourhood associations encouraged community members to surveil each other. However, the digitisation of citizens' information, massive data breach scandals, and expectations from the European Union prompted Japan to enforce stronger data privacy protections. In 2019, Japan enacted the Act on the Protection of Personal Information that created the world's largest safe data flow. Marcén compares this to how South Korea – another European Union trading partner – deals with the compliance demand. South Korea is more aggressive in safeguarding data privacy than Japan; it has established an independent governing body to enforce and supervise data protection laws even though it does not strictly enforce them. The comparison between the two countries shows that countries first changed their laws because of international laws, then they assimilated the concepts by bringing in their own cultural characteristics.

The next three chapters look at how governments regulate the development and consumption of media technologies. Governments exhibit conflicting attitudes towards new technologies: On the one hand, they understand that the digital media economy is a significant part of the economy; on the other, they fear that new technologies will break up social fabrics and ruin morality. This attitude is exhibited in the *Pokémon GO* phenomenon that swept across several countries in 2016. In Chapter 4, Deirdre Sneep examines how the Japanese government and media discussed augmented reality gaming in public places. She shows that the government issued warnings about accidents and health hazards caused by addictive handheld gaming. The media also disproportionately reported on accidents and deaths caused by drivers whose attention was diverted by playing video games. Anxieties about public health and safety as well as moral decline are nothing new. They were also expressed when the Sony Walkman became popular in Japan. Some of these anxieties can be explained by the different attitudes towards technologies between the older and the younger generations. While the former sees technologies as only for work, the latter sees them as playthings.

Intellectual property laws conceptualise media goods as abstract economic products, thus distributors and consumers of illegal media are seen to be motivated by economic reasons. As Micky Lee and Weiqi Zhang demonstrate in Chapter 5, North Koreans' illegal distribution and consumption of South Korean media and Hollywood films shows that the discussion of intellectual property violations does not take into account the local contexts in which these goods are distributed and consumed. They bring in the concepts of materiality and corporeality to show that material properties and bodies matter in the smuggling of media goods across borders, because the majority of the population does not have Internet access and the government considers possessing foreign media a crime. However, Lee and Zhang also caution that Western news reports about everyday life in North Korea aim to differentiate the East from the West. In these reports, North Koreans are portrayed as being animal-like, with foreign media being the saviour that can lift them up to become enlightened beings. The dichotomy between the East and West, between North Koreans and Western beings is problematic because news discourse sees capitalism as a superior system, thus attributing negative qualities to non-capitalist systems.

The North Korean government is also conflicted about new media technologies. As one of the most closed countries in the world, the government resists importing 'Western'-styled media technologies and content. Its impoverished economic status and lack of expertise also mean that the state cannot develop its own media technologies. However, image-saturated South Korean media have definitely influenced how North Korean leaders view the power of televisual images. Elizabeth Shim examines the nuclear weapon images released by the North Korean government. In Chapter 6, she argues that the North Korean government employs the simulacra of nuclear weapons in its propaganda campaign. Whether North Korea really possesses nuclear weapons that can destroy South Korea and Japan is not of primary concern; it understands how images of nuclear weapons can demonstrate its strength among the local populations and establish its international status.

Part III explores how government policies of building a digital economy reorganise labour. The case of South Korea is a good example of how the government accommodates surplus labour – mostly comprising young people – after a rapid period of economic development interspersed with financial meltdowns. On the surface, South Korea brands itself as a leading digital society with vibrant esports and video gaming scenes as well as enterprising online entrepreneurs. However, two chapters examine how this façade of a digital society covers up youth underemployment. In Chapter 7, Keung Yoon Bae looks at the export of South Korean esports gamers to non-Korean professional teams. She argues that the esports industry underwent a vertical integration when game developers began to invest in professional esports teams. South Korean players are believed to be better players because of the perceived innate abilities. However, it is also possible that working for an overseas professional team is a way to absorb the surplus of esports professionals in the country.

In Chapter 8, Peichi Chung analyses the esports industry by shifting a corporation-centred approach to an alternative approach that centres on the players and the policy model of innovation. Chung explores how the voices of various players can be contextualised in a larger societal structure. She draws her evidence from twenty in-depth interviews of esports professionals and shows the dilemmas they face in developing their careers. She found that South Korea dominates the esports scene in Northeast Asia not only because of two public organisations that have built national and global infrastructure, but also because of player-driven esports innovation. A lot of esports professionals dreamed of becoming professional players at the beginning of their careers.

When they failed to make pro, they developed value chains in game publishing, athletics, education, media, and regulation.

Keiji Amano and Geoffrey Rockwell examine the popular game pachinko, which is rarely found outside Japan, in Chapter 9. Amano and Rockwell argue that there are at least three phases of the pachinko industry in Japan; popular media such as films and novels reflect how filmmakers and novelists saw the game differently in these three phases. The first phase commenced after the end of World War II, at that time pachinko was a family-owned business and the game was seen as an idle pursuit, going against the work ethic expected of adults to help rebuild the country. The second phase came during the economic boom of the 1970s and 1980s, when the industry became more organised but also more opaque. Many believe that pachinko owners are gangsters or Koreans who were sympathetic to the North Korean government. During this period, pachinko parlours served as the glamorous backdrops for films. The third phase came in the post-economic bubble era, when pachinko was transformed from a mechanical technology to a digital one. Players' skills were no longer relevant, and the game became one of chance. The transformation of the technology serves as a metaphor for life's randomness; those who lose in this random game are like the populations who were left behind in the post-bubble economy.

In Chapter 10, Shawn Bender examines how the ageing population in Japan prompted robot manufacturers to consider producing robots for domestic uses, in particular caring for the elderly. From his ethnographical fieldwork, he found robotic engineers changed their understanding of robots. Initially, robots were in animal form, allowing patients to regain dignity by teasing them. Later, the robots acted like a digital social platform that helps forge social ties among the patients. He concluded that robots are not substitutes for humans, but facilitate human interactions through the material body of robots. As such, the robots are not designed for work, but for play; they help the elderly who have forgotten how to play connect with other humans.

In the Conclusion, we discuss the following five questions and suggest how the contributors answered them. First, how does digital technology change labour practices and industry structures in electronic gaming? Second, how does play foster subjectivity in a corporation-dominated digital environment? Third, how do analogue and digital technologies shape the meanings of work and play? Fourth, how do work and play in a local setting challenge abstract concepts such as intellectual property, data privacy, sociality, and a state-planned economy? Fifth, how are regions created through work and play of media contents and media ecosystems? We hope that this book will be able to provide readers a new understanding of media technologies for work and play in Japan and the Koreas.

References

Ahmed, A.K.N. (1988) 'Basis of Japan's modernization', *Economic and Political Weekly*, 23(33): 1674–6.

Asif, S. (2017) '10 percent of Samsung's revenue came from its home market South Korea', *Sammobile*, Available from: https://www.sammobile.com/2017/07/24/10-percent-of-samsungs-revenue-came-from-its-home-market-south-korea/ [Accessed 21 March 2020].

Baek, J. (2016) *North Korea's Hidden Revolution: How the Information Underground is Transforming a Closed Society*, New Haven, CT: Yale University Press.

Bossy, S. (2000) 'Academic pressure and impact on Japanese students', *McGill Journal of Education*, 35(1): 71–89.

Boutilier, A. (2015) 'Video game companies are collecting massive amounts of data about you', *The Star*, 29 December, Available from: https://www.thestar.com/news/canada/2015/12/29/how-much-data-are-video-games-collecting-about-you.html [Accessed 1 January 2021].

Burgess, M. (2016) 'North Korea has just 28 websites registered to its domain – and most celebrate Kim Jong-Un', *Wired*, 21 September, Available from: http://www.wired.co.uk/article/north-korea-28-websites-domain [Accessed 22 March 2020].

Chakraborty, A. (2018) 'Olympic a distant dream despite Asian boost: Esports chief', *Reuters*, 29 August, Available from: https://www.reuters.com/article/us-games-asia-esport/olympic-a-distant-dream-despite-asiad-boost-esports-chief-idUSKCN1LE13W [Accessed 25 February 2019].

Chalk, A. (2017) 'Faker's first Twitch stream sets a record for viewer', *PC Gamer*, 7 February, Available from: https://www.pcgamer.com/fakers-first-twitch-stream-sets-a-record-for-viewers [Accessed 25 February 2019].

Chan, D. (2013) 'Locating play: The situated localities of online and handheld gaming in East Asia', in J.A. Lent and L. Fitzsimmons (eds) *Asian Popular Culture: New, Hybrid and Alternative Media*, Lanham, MD: Lexington, pp 17–34.

Chen, L. (2018) 'Historical and cultural features of Confucianism in East Asia', in R.T. Ames and P.D. Hershock (eds) *Confucianisms for a Changing World Culture Order*, Honolulu, HI: University of Hawai'i Press, pp 102–11.

Ching, L. (2000) 'Globalizing the regional, regionalizing the global: Mass culture and Asianism in the age of late capital', *Public Culture*, 12(1): 233–57.

Choi, J.B. (2010) 'Of the East Asian cultural sphere: Theorizing cultural regionalization', *China Review*, 10(2): 109–36.

Choi, J.H-j. (2010) 'The city, self-connection: "Transyouth" and urban social networking in Seoul', in S.H. Donald, T.D. Anderson, and D. Spry (eds) *Youth, Society and Mobile Media in Asia*, London: Routledge, pp 88–107.

Chow, R. (2010) *The Rey Chow Reader*, New York: Columbia University Press.

Clark, D.N. (2012) *Korea in World History*, Ann Arbor, MI: The Association for Asian Studies.

Cleveland, M. and Bartsch, F. (2018) 'Global consumer culture: Epistemology and ontology', *International Marketing Review*, 36(4): 556–80.

Corrigan, T. (2015) 'Media and cultural industries internships: A thematic review and digital labour parallels', *TripleC: Communication, Capitalism & Critique*, 13(2): 336–50.

Craig, W. (2018) 'The nature of leadership in a flat organization', *Forbes*, 23 October, Available from: https://www.forbes.com/sites/williamcraig/2018/10/23/the-nature-of-leadership-in-a-flat-organization/#2c954f085fe1 [Accessed 1 January 2021].

Denyer, S. and Kim, M.J. (2019) 'In South Korea's hypercompetitive academia, esports gamers carve out larger niche', *Washington Post*, 23 August, Available from: https://www.washingtonpost.com/world/asia_pacific/in-south-koreas-hypercompetitive-academia-esports-gamers-carve-out-larger-niche/2019/08/22/e92dedba-8e08-11e9-b6f4-033356502dce_story.html [Accessed 1 January 2021].

Dodds, K. (2007) *Geopolitics: A Very Short Introduction*, New York: Oxford University Press.

Eitzen, D.S. and Zinn, M.B. (2012) *Globalization: The Transformation of Social Worlds*, Belmont, CA: Wadsworth Cengage Learning.

Esports Earnings (2019) 'Top 500 highest earnings for Korea, Republic of', Available from: https://www.esportsearnings.com/countries/kr-x400 [Accessed 1 January 2021].

Flanagan, M. and Nissenbaum, H. (2014) *Values at Play in Digital Games*, Cambridge, MA: The MIT Press.

Freedom House (2013) *Freedom of the Press 2013*, Washington DC: Freedom House.

Fuchs, C. (2012) 'Google capitalism', *TripleC: Communication, Capitalism & Critique*, 10(1), Available from: https://www.triple-c.at/index.php/tripleC/article/view/304 [Accessed 22 March 2020].

Gallagher, M. (2017) 'KakaoTalk meets the Ministry of Education: Mobile learning in South Korean higher education', in H. Crompton and J. Traxler (eds), *Mobile Learning and Higher Education: Challenges in Context*, New York: Routledge, pp 136–47.

Ge, S. (2010) 'How does Asia mean? (Part I)', *Inter-Asia Cultural Studies*, 1(1): 13–47.

Gillespie, T., Boczkowski, P.J., and Foot, K. (2014) (eds) *Media Technologies: Essays on Communication, Materiality, and Society*, Cambridge, MA: The MIT Press.

Harvey, D. (2007) *A Brief History of Neoliberalism*, New York: Oxford University Press.

Hastings, J. (2016) *A Most Enterprising Country: North Korea in the Global Economy*, Ithaca, NY: Cornell University Press.

Hemmert, M. (2007) 'The Korean innovation system: From industries catch-up to technological leadership?', in J. Mahlich and W. Pascha (eds), *Innovation and Technology in Korea*, New York: Springer, pp 11–32.

Hjorth, L. (2010) 'The piece of being mobile: Youth, gender and mobile media', in S.H. Donald, T.D. Anderson, and D. Spry (eds) *Youth, Society and Mobile Media in Asia*, London: Routledge, pp 73–87.

Hjorth, L. and Khoo, O. (2016) 'Intimate entanglements: New media in Asia', in L. Hjorth and O. Khoo (eds) *Routledge Handbook of New Media in Asia*, London: Routledge, pp 1–14.

Holroyd, C. and Coates, K. (2012) *Digital Media in East Asia: National Innovation and the Transformation of the Region*, Amherst, NY: Cambria.

Hu, E. (2015) 'The all-work, no-play culture of South Korean education', *NPR*, 15 April, Available from: https://www.npr.org/sections/parallels/2015/04/15/393939759/the-all-work-no-play-culture-of-south-korean-education [Accessed 1 January 2021].

Iwabuchi, K. (2002) *Recentering Globalization: Popular Culture and Japanese Transnationalism*, Durham, NC: Duke University Press.

Jin, D.Y. (2010) *Korea's Online Gaming Empire*, Cambridge, MA: The MIT Press.

Jin, D.Y. (2017) 'Construction of digital Korea: The evolution of new communication technologies in the 21st century', *Media, Culture & Society*, 39(5): 715–26.

Jun, J., LaFoy, S., and Sohn, E. (2015) *North Korea's Cyber Operations: Strategy and Responses*, Lanham, MD: Rowman & Littlefield.

Kaser, R. (2018), 'Olympics Committee President rejects esports as 'promoting violence'', *TNW*, 4 September, Available from: https://thenextweb.com/gaming/2018/09/04/olympics-esports-violence/ [Accessed 25 February 2019].

Kim, G. (2013) 'Mad-cow disease and alternative YouTube videos: Brechtian politics of aesthetics in grassroots media spectacles, voluntary mobilization, and collective governance from Korea's candlelight movements', in J.A. Lent and L. Fitzsimmons (eds) *Asian Popular Culture: New, Hybrid and Alternative Media*, Lanham, MD: Lexington, pp 109–43.

Kim, J.E. (2015) 'New media and new technology in colonial Korea: Radio', in A. Prescott (ed) *East Asia in the World: An Introduction*, New York: Routledge, pp 140–53.

Kim, J.Y. (2014) 'South Korea and the sub-empire of anime: Kinesthetics of subcontracted animation production', *Mechademia*, 9: 90–103.

Kim, Y.H. (2004) 'North Korea's cyberpath', *Asian Perspective*, 28(3): 191–209.

Kim, Y-S. (1996) 'Korea and US industry-technology cooperation', *The Journal of East Asian Affairs*, 10(1): 1–12.

Kiyohara, S., Maeshima, K., and Owen, D. (2018) (eds) *Internet Election Campaigns in the United States, Japan, South Korea, and Taiwan*, Cham, Switzerland: Palgrave Macmillan.

Kline, S. and de Peuter, G. (2002) 'Ghosts in the machine: Postmodern childhood, video gaming and advertising', in D. Cook (ed) *Symbolic Childhood: Popular Culture and Everyday Life*, New York: Peter Lang, pp 255–78.

Ko, K., Lee, H., and Jang, S. (2009) 'The Internet dilemma and control policy: Political and economic implications of the Internet in North Korea', *The Korean Journal of Defense Analysis*, 21(3): 279–96.

Lai, S. (2018) 'Countries with the fastest internet in the world 2018', *Atlas and Boots* [Blog], 30 September, Available from: https://www.atlasandboots.com/remote-jobs/countries-with-the-fastest-internet-in-the-world [Accessed 25 February 2019].

Larson, J.F. (1995) *The Telecommunications Revolution in Korea*, Hong Kong: Oxford University Press.

Lee, D-h. (2012) '"In bed with the iPhone": The iPhone and hypersociality in Korea', in L. Hjorth, J. Burgess, and I. Richardson (eds) *Studying Mobile Media: Cultural Technologies, Mobile Communication, and the iPhone*, New York: Routledge, pp 63–81.

Lincoln, J.R. and Kalleberg, A.L. (1985) 'Work organization and workforce commitment: A study of plants and employees in the U.S. and Japan', *American Sociological Review*, 50(6): 738–60.

Martin, T.W. and Chomchuen, W. (2017) 'North Koreans get smartphones, and the regime keeps tabs', *Wall Street Journal*, 6 December, Available from: https://www.wsj.com/articles/north-koreans-get-smartphones-and-the-regime-keeps-tabs-1512556200 [Accessed 1 January 2021].

Matsuda, M. (2010) 'Japanese mobile youth in the 2000', in S.H. Donald, T.D. Anderson, and D. Spry (eds) *Youth, Society and Mobile Media in Asia*, London: Routledge, pp 31–42.

McRobbie, A. (2016) *Be Creative: Making a Living in the New Culture Industries*, Cambridge: Polity.

Ministry of Economy, Trade, and Industry (2012) 'Cool Japan Strategy', Available from: https://www.meti.go.jp/english/policy/mono_info_service/creative_industries/creative_industries.html [Accessed 1 January 2021].

Newzoo (2018) '2018 Global Esports Market Report', Available from: https://newzoo.com/insights/trend-reports/global-esports-market-report-2018-light/ [Accessed 1 January 2021].

Nichols, R. (2014) *The Video Game Industry*, London: Bloomsbury.

Nintendo (n.d.) 'Financial highlights', Available from: https://www.nintendo.co.jp/ir/en/finance/highlight/index.html [Accessed 29 October 2019].

Nishioka, Y. and Sugaya, M. (2014) 'Japan's legislative framework for telecommunications: Evolution toward convergence of communications and broadcasting', in Y-l. Liu and R.G. Picard (eds) *Policy and Marketing Strategies for Digital Media*, New York: Routledge, pp 120–37.

Noland, M. (2008) *Telecommunications in North Korea: Has Orascom Made the Connections?*, Washington, DC: Peterson Institute for International Economics and East-West Center.

Oh, M. and Larson, J.F. (2011) *Digital Development in Korea: Building an Information Society*, London: Routledge.

Otmazgin, N. (2013) 'Popular culture and regionalization in East and Southeast Asia', in N. Otmazgin and E. Ben-Ari (eds) *Popular Culture Co-productions and Collaborations in East and Southeast Asia*, Singapore: National University of Singapore Press, pp 29–51.

Park, M. and Curran, J. (2000) *De-westernizing Media Studies*, London: Routledge.

Picard, M. (2013) 'The foundation of geemu: A brief history of early Japanese video game', *Game Studies*, 13(2), Available from: http://gamestudies.org/1302/articles/picard [Accessed 22 March 2020].

Pile, K. (1996) *The Making of Modern Japan* (2nd edn), Lexington, MA: D.C. Heath.

Saito, K. (2013) 'Regionalism in the era of eco-nationalism: Japanese landscape in the background art of games and anime from the late-1990s to the present', in J.A. Lent and L. Fitzsimmons (eds) *Asian Popular Culture: New, Hybrid and Alternative Media*, Lanham, MD: Lexington, pp 35–58.

Scholz, T. (2013) (ed) *Digital Labor: The Internet as Playground and Factory*, New York: Routledge.

Shirahata, Y. (1998) 'The modernization of transport and communications in Japan', in F. Günergun and S. Kuriyama (eds) *The Introduction of Modern Science and Technology to Turkey and Japan*, Tokyo: International Research Center for Japanese Studies, pp 75–86, Available from: http://doi.org/10.15055/00001603 [Accessed 1 January 2021].

Spry, D. (2010) 'Angels and devils: Youth mobile media politics, fear, hope and policy in Japan and Australia', in S.H. Donald, T.D. Anderson, and D. Spry (eds) *Youth, Society and Mobile Media in Asia*, London: Routledge, pp 15–30.

Sternberg, R. (2003) 'New media policies and regional development in Japan', in H-J. Braczyk, G. Fuchs and H-G. Wolf (eds) *Multimedia and Regional Economic Restructuring*, London: Routledge, pp 346–75.

The Telegraph (2018) 'Inside North Korea: Photographers offer a rare insight into Kim Jong-un's secretive state', 5 June, Available from: https://www.telegraph.co.uk/news/0/inside-north-korea-photographer-offers-rare-insight-kim-jong/people-read-communal-newspaper-pyongyang-subway-system-claimed/ [Accessed 1 January 2021].

Tu, W. (2000) 'Implications of the rise of "Confucian" East Asia', *Daedalus*, 129(1): 195–218.

Uchida, J. (2012) *Brokers of Empires: Japanese Settler Colonialism in Korea: 1876–1945*, Cambridge, MA: Harvard University Press.

Wajcman, J. (1991) *Feminist Confronts Technology*, University Park: Pennsylvania State University.

Yoo, E. and Lee, H. (2014) 'The impact of digital convergence on the regulation of new media in Korea: Major issues in new media policy', in Y-l. Liu and R.G. Picard (eds) *Policy and Marketing Strategies for Digital Media*, New York: Routledge, pp 138–53.

Yoon, K. (2010) 'The representation of mobile youth in the post-colonial techno-nation of Korea', in S.H. Donald, T.D. Anderson, and D. Spry (eds) *Youth, Society and Mobile Media in Asia*, London: Routledge, pp 108–19.

Zhang, W. and Lee, M. (2019) 'Black markets, red states: Media piracy in China and the Korean Wave in North Korea', in Y. Kim (ed) *South Korean Popular Culture and North Korea*, New York: Routledge, pp 83–95.

Ziccardi, G. (2013) *Resistance, Liberation Technology and Human Rights in the Digital Age*, Dordrecht: Springer.

PART I

Gender Online and Digital Sex

Introduction to Part I

Micky Lee

The two chapters in Part I discuss how sex and gender are understood in a digital environment in Japan and South Korea. In the first chapter, Kyooeun Jang discusses how online commerce appears to create new business opportunities for young women in South Korea, but the strong push for female entrepreneurship is actually a tactic to solve the problem of mass youth unemployment. Young women internalise a belief that if they do not achieve commercial success, it is because they are not passionate enough for their business. In the online environment, young women also do emotional labour, exploit their sexuality, and balance work and family responsibilities without knowing about the structural issues at play obstructing their finding gainful employment.

In the second chapter, Peter Galbraith examines a niche video game genre – adult-themed PC games – and questions whether it is possible to draw a boundary between Japan and the rest of the world in the digital globalised era. Although this video game niche only has a small number of players, its rape themes have ignited international condemnation. In order to avoid the spotlight, Japanese gamers avoid international attention by limiting the geolinguistic area within which these games are played. However, some gamers believe the survival of this niche depends on non-Japanese gamers outside the country.

The chapters in this part contribute to a discussion in digital media studies by asking whether new information and communication technologies would alter gender relations or reinforce them. In addition, they show how historical gender relations in Japan and South Korea shape how gender is experienced. As emphasised in the introduction, we do not aim to compare East Asia with the

West. We believe gender relations are shaped by local practices of digital technologies in different countries. We first summarise some major schools of thought that concern women, gender, and digital technologies.

Since the advancement of digital technologies in the 1990s, there has been a debate about whether technologies would change gender relations or reinforce gender inequality. The Internet and mobile devices are believed to simultaneously liberate and oppress women. On the one hand, digital technologies allow women to participate in civic, public, and economic life. On the other hand, online sexual harassment is believed to oppress women, further exploiting them as sexual beings (Broadband Commission for Digital Development, 2015).

On the positive side, digital technologies may bring unprecedented opportunities for women, especially those who are traditionally barred from information because of illiteracy or restrictive local customs. Technologies are believed to increase women's access to information, helping them make better decisions for themselves and their families. The Internet is also believed to increase women's purchasing power as they can shop online and compare prices. Lastly, women are encouraged to study computer science because technical knowledge will help them find high-paying jobs (ITU, n.d.).

Despite the opportunities, digital technologies also have their negative side. The anonymity in the online world exacerbates sexual harassment, sexual exploitation, and gender bias. This is particularly the case in countries where regulations are more relaxed for the Internet than for mass media.

The debate on whether digital technologies improve women's status or reinforce inequality has been criticised for being overly simplistic. This debate relies on the assumption that technologies are tools and that women are a fixed category. This criticism can be broken down in three areas. First, the earliest studies tend to compare how women use digital technologies differently from men. The assumption is that men's use of digital technologies is the norm that women should follow. It is further assumed that once women use technologies more like men, gender equality will be achieved. This vision of gender equality does not acknowledge that women may use technologies differently because they make different meanings of technology and they use them for different purposes (Wajcman, 1991).

Second, the earliest studies assumed a male–female gender binary and do not acknowledge that there may be populations who do not identify with either gender. In other words, the earliest studies assume biological sex determines gender identity, and users cannot experience a different identity online. Precisely because the Internet allows anonymity, users can experience gender online in a way that they cannot in real life. Therefore, scholars such as Butler (1999) suggest that gender is being performed through social actions: We perform our gender through talking and acting in social situations. The Internet offers a good performing space for social beings to enact their gender.

Third, the earliest studies see technologies as tools that women can harness if they want to improve their political, social, and economic statuses. However, science and technology studies scholars (Gillespie et al, 2014) argue that technologies are more than tools. Like media content, technologies are texts of which meanings are made based on users' social background and cultural understanding. In addition, political economists argue that digital technologies are commodities that transnational corporations (such as Microsoft, Google, and Apple) produce to increase profits. Therefore, merely using technologies will not advance women, but will instead draw them closer to a global economy that further exploits women's labour and commodifies gender relations (Lee, 2006). For example, transnational corporations such as Apple set up factories in countries with cheaper labour cost, pushing women to leave their villages to work in cities to supplement their family income. From the consumption side, digital technologies are advertised as ways to make up for lost time with loved ones. For example, mothers who are busy working outside the home can chat with their children on the phone to compensate for their absence.

After considering how the relations between women, gender, and technologies can be studied, we now ask how these studies can apply to the cases in East Asia. Before answering this question, we first show some statistics about women's status in Japan and South Korea. After that, we suggest how the two chapters in this part bring in a gender dimension to understanding women's status in these two East Asian countries.

According to UN Statistics (United Nations, 2019), girls and boys in Japan receive similar levels of education at the secondary level. However, in their adulthood, men and women are not equal in receiving higher education, working outside the home or running for public office. In Japan, women are less likely to receive higher education and few are admitted to medical schools (Shirakawa, 2019). Males over the age of 15 are much more likely to participate in the labour force than women, even though men's participation rate has been dropping while women's has been increasing. In 2017, only 51 per cent of women over the age of 15 worked for pay outside the home. Unsurprisingly, women spend more time working on domestic chores and unpaid care work. Politically, only 38 per cent of national parliament seats are occupied by women. A similar situation prevails in South Korea. Girls are as likely as boys to have a secondary education, but women are less likely to work outside home for pay. In 2017, only 53 per cent of women above the age of 15 worked outside the home, even though the participation rate has been increasing. South Korean women also spend more time doing unpaid domestic work and care work. Korean women, when compared to Japanese women, are even less likely to occupy a seat in the national parliament; only 17 per cent of the seats are held by women. In both countries, girls are given equal opportunities to participate in public life until they reach adulthood, when women are less likely to work outside the home for pay or run for public office. Women are expected to spend more time working on household chores.

The Global Gender Gap Report 2020 published by the World Economic Forum further confirmed the low participation rate in politics and the economy among Japanese and South Korean women. In terms of women's economic participation and opportunity, Japan and South Korea are ranked at 115th and 127th respectively among some 150 countries. In terms of political empowerment of women, Japan and South Korea are ranked 144th and 79th respectively. In both indices, the two countries lag behind other Asian countries that are considered less economically powerful, such as Laos, Thailand, Cambodia, Vietnam, Bangladesh, and India.

Women in Japan and South Korea then experience a paradox: On the one hand, they live in economic prosperity, enjoy political freedom, and receive universal education; on the other hand, they do not fully participate in public life by working outside the home or running for public office. Despite the huge economic and political strides that Japan and South Korea have made since World War II, women have not gained much real economic and political power.

In addition, the collectivist culture in the two societies pushes people to follow group norms, such as gender expectations and expressions. It is harder for individuals to experiment with gender in social or public settings, because members of society are expected to fulfil prescribed gender roles. Besides, gender relations are still highly patriarchal in Japan and South Korea: Men are still in dominant power positions, that let them decide resource allocation. How can the issues of gender online and digital sex be understood in societies that have collectivist cultures and patriarchal social relations?

The two chapters in this part show how an online space enables users to experience different gender expressions and challenge gender relations through e-commerce and digital gaming. At the same time, they also point out that gender expressions and gender relations in the offline space constrain gender experience in the online space. In the first chapter, online entrepreneurship appears to be a new space for South Korean women to engage in electronic commerce. As only one in two Korean women works for pay outside home, online entrepreneurship appears to be an ideal space where they can earn an income from home. Despite the rosy promise, Kyooeun Jang shows that women perform considerable emotional labour in the online space because they are expected to act like friends to their customers. Online entrepreneurs are also expected to be responsive to customers' demands. At the same time, they cannot stand out among their customers or else they will be deemed to be too showy. In this space, they also have to conform to gender norms and expectations that are upheld by men in a patriarchal society. For example, female entrepreneurs are expected to see themselves as sexualised beings even though the majority of their customers are women. Female entrepreneurs need to please other women who exercise the male gaze on them. In addition, mother-entrepreneurs are expected to show devotion to the families, performing the role of good mothers, even though they need to constantly be present online and connected with their followers.

Adult PC games are seen as quintessential Japanese genres that are so culturally specific that they cannot successfully find an audience outside the country. However, despite having only a niche consumer demographic, adult PC games have caused an international uproar. Themes such as raping young girls in public places have caused many international communities to pressure the Japanese government to stamp out the genre. To respond to this, adult PC game developers and gamers hope to 'contain' the genre within the country by disallowing non-Japanese members to attend their events.

This case shows how gender relations can be regulated within and outside a country's borders. The international community is alarmed that the Japanese government allows the sale within and outside the country of video games that depict harm to young girls. Coupled with the low political and economic participation rate among Japanese women, adult PC games seem to reinforce the 'backwardness' of gender relations in Japan. However, some gamers argue that despite the rape themes, Japan has a low rate of sexual assault. This comment reflects an enduring debate on whether media representations lead to real-life behaviours or whether the audience understands that media representations have no bearing on real-world social relations. While this question is not addressed in this chapter, the case of adult PC games questions whether the international community can regulate gender relations and whether gender relations can be contained within a national border.

The international community believes that stamping out this genre will protect young girls in the country and elsewhere. The assumption is that gender relations are fixed and unchanging: Male gamers are predators of young girls; all adult PC game players are heterosexual males who lust after young girls. Is it possible that some players are females and some others prefer same-sex relationships? For those who want such games to be contained in Japan, they seem to suggest that the international community has no business interfering with how gender relations are represented in a Japanese genre and that any consumption of such games by non-Japanese users will taint Japanese culture. In this argument, gender relations are also unchanged because what marks Japan as different from other cultures is how gender is represented in this game genre.

References

Broadband Commission for Digital Development (2015) *Cyber Violence against Women and Girls: A World-wide Wake-up Call*, Geneva and Paris: ITU and UNESCO.

Butler, J. (1999) *Gender Trouble: Feminism and the Subversion of Identity*, New York: Routledge.

Gillespie, T., Boczkowski, P.J., and Foot, K. (eds) (2014) *Media Technologies: Essays on Communication, Materiality, and Society*, Cambridge, MA: The MIT Press.

ITU (International Telecommunications Union) (n.d.) 'Why a girls in ICT day?', Available from: https://www.itu.int/en/ITU-D/Digital-Inclusion/Women-and-Girls/Girls-in-ICT-Portal/Pages/Why-a-Girls-in-ICT-Day.aspx [Accessed 1 January 2021].

Lee, M. (2006) 'What's missing in feminist research in new information and communication technologies?', *Feminist Media Studies*, 6(2): 191–210.

Shirakawa, T. (2019) 'Japan an undeveloped country for women', *The Japan Times*, 11 January, Available from: https://www.japantimes.co.jp/opinion/2019/01/11/commentary/japan-commentary/japan-underdeveloped-country-women/#.Xl-upkpOmUk [Accessed 22 March 2020].

United Nations (2019) 'Minimum set of gender indicators', Available from: https://genderstats.un.org/#/home [Accessed 1 January 2021].

Wajcman, J. (1991) *Feminism Confronts Technology,* State College: Pennsylvania State University Press.

World Economic Forum (2020) 'Global Gender Gap Report 2020', Geneva: World Economic Forum.

Sharing, Selling, Striving: The Gendered Labour of Female Social Entrepreneurship in South Korea

Kyooeun Jang

Viewing the digital economy through rose-coloured lenses

Entering any municipal-funded community centre (also called 'cultural centres') in South Korea, one can easily find advertisements encouraging citizens to learn a new musical instrument, sport, or language at affordable prices. In recent years, however, the bulletin boards have been patched with new flyers that read: 'Achieve your entrepreneurial dream at our new female entrepreneur platform! One-on-one consulting sessions' or 'Mompreneur business academy: Grand opening!' specifically targeting young mothers. In the private sector, companies such as Facebook Korea have partnered with Korea's leading women's universities to implement internship opportunities to nurture female entrepreneurs (Cho, 2016) and have launched the '#shemeansbusiness' programme that offers digital marketing and networking tips to female business owners (Kim, 2016). The increasing number of millennial women engaging in self-employment through online micro-entrepreneurship has been praised by governments and corporations around the world as a 'movement away from bureaucratic, male-dominated work structures' through which women can enjoy lifestyles 'where passion and profit meld' (Duffy and Pruchniewska, 2017: 844). Women have been especially impacted by the growing trend of 'working for yourself' (Taylor, 2015: 176), as they are enticed by the flexibility of working independently, a situation that helps them manage the dual roles of full-time employee and caretaker at home (177).

South Korea boasts of its leadership in successfully integrating the digital into governance, education and daily life, and has recently implemented various governmental and corporate initiatives encouraging female labour participation through the use of digital media technologies. Yet, it ranks 118 out of 144 countries in terms of women's 'economic participation and opportunity' and 'political empowerment', making it one of the lowest-scoring countries in Asia and well below the global average, ranking lower than countries such as Cambodia and Senegal (World Economic Forum, 2017: 11). Women hold a meagre 2 per cent of senior management positions, which is only one tenth of the average of countries in the Organisation for Economic Co-operation and Development. As women face challenges being rehired at corporations after leaving work for childbirth in their 30s, many miss 'a decade or more of prime working life' and resort to part-time jobs, never reaching 'the next rung on the career ladder' (Lagarde, 2017: para. 12). In this context, online micro-entrepreneurship is seen by many Korean women as the answer to secure employment, as working for oneself is seen as a way to successfully combine profits, passions, and, in the case of young mothers, parenthood.

A more comprehensive understanding of this phenomenon enabled by a changing relationship of work, play, and digital media technologies, therefore, requires an exploration of relevant labour practices and gendered performances that evolve alongside online work. While it is believed that access to cutting-edge technological infrastructure could lift barriers to gender inequality in terms of Internet access and digital literacy, gendered social norms continue spilling over into the virtual space and may hinder women from fully realising their professional potential. In this sense, South Korea is an ideal site for examining how feminist analyses of technology are shifting from preliminary concerns over 'women's access to technology' (Wajcman, 2010: 146) into how gender and work are constructed through the socio-cultural practices in which technology is utilised and monetised.

After first contextualising the rise of micro-entrepreneurship within the larger economic context of how digital media technologies are reconceptualising work and play, this chapter analyses performances of gender on Instagram business accounts managed by women micro-entrepreneurs engaged in fashion retail and lifestyle services. Many of these women, who grew up as 'digital natives' by using Internet communications technologies from early childhood, are among the first to build careers in the digital industries, which require employees to engage in 'creative labour'. Creative labour deals with the production of 'nonmaterial goods' (Hirsch, 1972: 642) that are 'primarily aesthetic and/or symbolic-expressive, rather than utilitarian and functional' in industries such as advertising, art, fashion, and communications technologies (Conor et al, 2015: 3; see also Introduction to Part III).

Additionally, through textual analysis of female micro-entrepreneurs' online business accounts and interviews with actual entrepreneurs, this chapter challenges the post-feminist discourse garnering mass appeal among millennial women, especially in South Korea. Post-feminism celebrates a woman's 'freedom of choice' in all areas of life, such as parenting, professional opportunities, and, especially, sexual empowerment (Negra and Tasker, 2007: 2). The idea stipulates that feminism has become irrelevant to society because gender equality has been achieved (Negra and Tasker, 2007: 8). Operating under an assumption that women have obtained 'full economic freedom', post-feminism suggests that a woman's retreat from the public sphere is purely a matter of individual choice and that today's economy and technologies have allowed for a system of pure meritocracy that allows women to fully 'compete in education and work' alongside men (Negra and Tasker, 2007: 32). This rhetoric, born of a neo-capitalist belief that views success as an individual feat rather than a result of social structures, overlooks the existing 'structural, cultural forces', including the 'lingering gender norms' that still affect the modern workplace (Duffy and Pruchniewska, 2017: 856). In this context, an exploration of the intersections of 'post feminism, neoliberal economies and social media labour' can illustrate how new types of work enabled by new technology may 'both conceal and reveal enduring structures of power' (Duffy and Pruchniewska, 2017: 856), forces easy to overlook in an era that glamorises technological development, entrepreneurship, and individual success.

Methods

This chapter draws on a mixture of qualitative methods consisting of face-to-face narrative interviews and online media textual analysis. Insights are drawn from female interviewees aged 25 to 30 who identify as independent entrepreneurs or marketers. These individuals operate their businesses using major social media platforms, particularly Instagram and Facebook, which are used by 51.3 per cent and 67.8 per cent of Korean Internet users respectively (Go, 2018), and Naver, South Korea's biggest search portal used by over 70 per cent of Korean Internet users (Hong, 2018). Interview questions requested respondents to speak about why they ventured into online micro-entrepreneurship, their day-to-day experiences as online entrepreneurs, and new skills they acquired through the job. Interviewees were also asked about their thoughts on whether their services and goods influenced ideals of gender among South Korean youth, and whether they needed to engage in gendered performances to achieve success online.

To supplement interview narratives and to visualise the working environments and practices, this study separately analysed twenty-two Instagram accounts that do not belong to the interviewees but are owned by well-known independent female micro-entrepreneurs involved in a variety of businesses, including fitness, nutrition, women's clothing, culinary art, and floral design. Each account (at the time of the study in 2017) had more than 10,000 followers. In their bio sections, the accounts explicitly mentioned that the sites were being used for e-commerce purposes. These accounts were chosen in accordance with the algorithms of the researcher's personal feed and her peer network's 'liked' content. Given the fact that about 40–50 per cent of the researcher's personal network comprises Korean women om their mid-20s to early 30s from different socio-economic backgrounds, the chosen accounts can be seen as being adequately representative of what Korean women in their 20s and 30s see on a daily basis on Instagram. Other accounts were selected from algorithmic recommendations that automatically pop up after clicking the 'follow' button on an entrepreneur's profile. This function provided a larger range of women in similar businesses, allowing for a comparison of self-curation practices among different women entrepreneurs in the same or similar industries. All content analysed here is publicly accessible. An Instagram account is needed to view public accounts both on the mobile app and via desktop.

New modes of production and labour: Immaterial labour

In today's 'post industrial service economy', goods and services are no longer restricted to material goods, but include 'immaterial goods' such as 'a service, a cultural knowledge or communication' (Wilkie, 2011: 82) as well as the very 'developments in communications and production technologies' (Wilkie, 2011: 88). In the past, if tools 'corresponded to different activities' such as a tailor's sewing machine or a weaver's loom, today, the computer has emerged as the 'universal tool […] through which all activities' are processed. This 'computerisation of production' has made 'abstract labour' more important in economic production (Hardt and Negri, 2001: 292), as the 'quality and nature of labour' must now incorporate 'information and communication' as foundational elements of work (Hardt and Negri, 2001: 289).

Abstract forms of labour, however, extend beyond services provided by the computer and also include work requiring human contact and care, such as 'service, knowledge, or communication' (Abidin, 2016: 89). This form of labour, also dubbed 'affective labour', includes activities that were previously not recognised as productive work, such as providing 'care, psychological support, and communication' (Arcy, 2016: 366) that involves human contact and attention (see Chapter 10). Affective labour is a 'labour in the bodily mode' (Hardt and Negri, 2001: 293) that generally refers to the 'ability to manage and distribute emotions' (Arcy, 2016: 367) which in turn contribute to the 'cultural content' of a commodity or brand (Arcy, 2016: 365).

This emotional work required in immaterial labour largely comprises the unpaid, domestic, and emotional endeavours that have been 'disproportionately' and historically assigned to women (Arcy, 2016: 366). Immaterial labour takes into account not only work in the form of information and knowledge, but also gendered performances that women must display to become successful. Such performances include visibility labour in which women entrepreneurs must make themselves attractive and reputable online. It also requires maintaining personal, affective relations fuelled by passion, hope, and community-building. Emphasis on so-called feminine soft skills, however, risks circumscribing women back into a traditional gender dichotomy that dictates males as producers and females as consumers or producers of underpaid labour.

Passion, play, and profits

Instagram, a compound word of 'instant' and 'telegram', was the brainchild of Bay Area engineers in 2010 who sought to build a service that allowed users to share photos and locations. The photo editing options including filters and frames made the app popular among American celebrities, and eventually caught the attention of major investors such as Mark Zuckerberg, who acquired the company in 2012 (Sengupta et al, 2012). In September 2015, Instagram began to monetise its platform with the global launch of a 30-second advertisement for advertisers (Finn, 2015) who previously could only market products via celebrity testimonials or organic reach. Since then, Instagram has become one of the most important social media e-commerce platforms worldwide, with over 25 million business accounts that are followed by 80 per cent of all users (Cohen, 2017).

Since Instagram prohibits hyperlinks from being placed within posts, many marketers use the platform for mostly branding and advertising purposes. They use other platforms to carry out monetary transactions and to provide more detailed product information. In Korea, Naver, which has expanded to host its own blogosphere, marketplace, and even GPS maps, is the social e-commerce platform of choice. Naver offers business owners free online space to host their shops, provided they conduct all transactions via 'Naver Pay', a platform-specific payment system favoured by users for its ability to simplify online transactions (Kim, 2017). However, because Naver charges an extra transaction fee from the business if a customer enters the business from Naver's main shopping portal and makes a purchase, many businesses are opting to advertise their Naver shops on third-party social media platforms such as Facebook or Instagram (Kim, 2018). Despite such complexities, more small- and medium-sized businesses in Korea are recognising social media platforms as an 'additional distribution channel' to increase profit (Han and Kim, 2016: 25) because they utilise users' existing personal networks, a feature that persuades users to 'perceive social media e-commerce as [a] more reliable' form of online commerce (Han and Kim, 2016: 27).

In this context, Korean youths are increasingly seeking to launch their own 'shopping mall' businesses (Lee, 2017a), a Konglish term that refers to social media e-commerce businesses specialising in fashion and beauty retail. Social media e-commerce has garnered exponentially increasing interest among young Korean women, who are enticed by the rags-to-riches stories of female 'shopping mall CEOs' who have become multi-millionaires despite having started with relatively minimal capital, and little professional education and experience compared to those running traditional brick-and-mortar businesses or even online businesses that are not based on social media. In Korean media, women such as Soo-kyung Kim, a successful blogshop entrepreneur who left school at age 17 to start a women's fashion retail business to support her family after the failure of her father's business, are hailed on TV shows and in news articles as role models and good daughters caring for their struggling families (Kwon, 2017). Stylenanda, one of the most famous woman-owned shopping malls in

Korea, is frequently cited as one of the most successful shopping malls that grew from a small online shop into a global brand that can now be found in department stores across Korea, Japan and even the US, and is admired by many aspiring fashion e-commerce entrepreneurs in Korea (Min and Jung, 2017).

However, the increasing number of young female entrepreneurs actually reflects a bleaker reality of the job market for Korean youths. The steady increase of female business owners in the past few years occurred alongside the exacerbation of youth unemployment (Choi, 2016). In 2017, only 7 per cent of Korean youths from ages 18 to 29 were employed in full-time jobs with health insurance and career prospects, and 70 per cent worked at temporary or contingent positions (Seoul Institute, 2017). For some, opening up small-scale online businesses is an inevitable choice to make ends meet, especially among many Korean youths and young mothers who struggle with finding regular, full-time employment. The number of 'career-foreshortened women' (경력 단절녀) managing online businesses has also increased, a trend that echoes the larger structural discrimination that discourages Korean women from entering the workforce after childbirth (Choi, 2016).[1] Due to the pervasiveness of a corporate culture that considers hiring women an 'unreliable bet' because female employees are seen as likely to leave after marriage and 'certainly after childbirth', women who do enter the workforce after marriage or raising children are most likely to end up in part-time and short-term jobs, a cycle that exacerbates the gender wage gap even further. Such hardships stem from conservative social norms that view women as being responsible for domestic labour regardless of employment status, as well as a lack of affordable childcare options (Draudt, 2016).

New currencies of aspiration, reputation, and attention

Hopeful beginnings: Aspirational costs and reputational investments

Kang, a female entrepreneur who owns a shopping mall selling clothing and accessories targeted at women in their 20s on Instagram and Naver, claimed that the hardships she was experiencing as a small-scale fashion entrepreneur could be overcome through self-discipline and perseverance. She cited as her role model the now-defunct global fast fashion brand Forever 21, which originated as a modest 900 square foot mom-and-pop shop owned by struggling Korean immigrants in California and grew into a multi-billion-dollar enterprise (Elkins, 2015).

I am doing this right now to support my family and because this is making me money. If this becomes a real brand, however, then things will be different. My social status will go up. The society we live in now idolises money over everything else. There is no need for honour or good education these days. The people with money hire educated people […] In my parents' era, studying gave you a way out, but not anymore […] They say shopping malls must suffer three years of losses before they really take off. I [am lucky because I] currently receive samples from my supplier for free, and the clothes I bought for personal use over the years serve as samples too. I even do private English tutoring on the side [which is why] I am not experiencing any financial losses […] I think I am at a much better position than others even though they may not think I am doing well. It's all about mind control.

Kang's account shows how profit is generated not only via sales but also through passion and perseverance. An aspirational and self-critiquing mindset helps Kang accept her present financial and emotional duress. During the months she fails to make a profit, she blames her lack of passion, a belief illuminating how such forms of flexible work persuade workers that their work solely 'serves the self and not the marketplace' (Tokumitsu, 2014). Even though young women increasingly recognise flexible online labour as a legitimate form of 'professional labour', a mentality of normalising present financial suffering for unguaranteed future profits may lead women to rationalise potentially exploitative work conditions. Furthermore, the phenomenon of young women labouring with a mindset fuelled by aspiration and a willingness to accept losses may exacerbate prejudices that view 'women as social sharers' (Duffy, 2015: 454) and as 'willing to work for social currency' such as recognition and approval, rather than for monetary wages, 'all in the name of love' (Duffy, 2015: 453).

Self-curation and self-promotion are crucial for women to become successful entrepreneurs on Instagram, a platform where consumers' perception of intimacy and attention directly translate into monetary profit. The very business model of Facebook and Instagram relies on ads for products that are sold based on the number of likes, shares, clicks, and even views, interactions originally used to express intimacy and attention within users' organic, social, and personal domains. In this sense, the digital economy is also a 'digital reputation economy' (Duffy, 2015: 454) in which 'reputation becomes a key commodity'

(Conor et al, 2015: 10) that requires continuous 'affective practices' (Duffy, 2015: 450) of networking, customer support, and emotional duress that ultimately become commodified and 'traded within capitalist systems of exchange' (Arcy, 2016: 367).

Min, a marketer at an art rental startup who partners with female lifestyle influencers on Instagram whose accounts are used as the primary platform for advertising Min's company, explains how the numbers of followers and likes have become new ways of measuring self-value in the online marketplace:

> We buy metrics that can quantify individuals, usually female lifestyle influencers, and we have different prices for those factors, like one's number of followers [...] a person's value is already given a price. Celebrities are valued in accordance to their filmography, fanbases or influence, and the value of those factors decide how much they earn when shooting ads. Likewise, our influencers are rated according to their lifestyle, Instagram account, personal exposure. We rate them as an A, B, C, or F. There are criteria for each level: reach, engagement, influence, and likes. I hire and fire these influencers based on this rate table. If one gets an F, it is like at school – they are fired. I keep score of each influencer [...] The rate changes by [the] hour [...] If they lose their value, we fire them [... because] to us, they are just money. We have a list of 100 to 120 influencers, and we continuously replace them. The women we fire usually don't show growth in followers or lack style, and ultimately fall behind.

Likewise, Kang reveals how she strives to maintain a favourable reputation by remaining modest and relatable to her mostly female audience. She sees her ability to read and manage women's emotions as crucial for her survival in the online fashion retail ecosystem:

> I know of one CEO who posted photos of her wearing clothes she bought from Dongdaemun Wholesale Market,[2] but paired with luxury handbags. Then obviously the viewers would feel distanced from her right? So she only received negative comments like 'I don't want to buy from her because I feel like I am just paying for her luxury bags'. She could have paired the clothes with cheap accessories from Dongdaemun, but she wanted to look

pretty and brag on SNS. [She was] acting snobby in an Audi convertible even though she was wearing cheap clothes from Dongdaemun [...] At first consumers thought she was pretty, but then, her bragging annoyed them. Since I am doing business with female customers, I have to be careful. I have a narrow waist, so when I sell an outfit I could potentially say, 'I am an extra small. Because I have a slim waist, these pants are loose-fitting.' But if I word it like that, [people will respond] 'What the hell? She is bragging about her figure. I'm not going to buy that!' So I speak in the third person. 'When a 159-cm model who wears XS wears [these pants], the waist is a bit loose' [...] I never say the model is me. I always refer to me as 'the model'. If people see that a professional model is wearing the clothes, they are more understanding, and just think, 'She is a model, so of course she has to maintain a nice body'. But when these women realise the model is also the CEO, they become jealous, you know?

Similarly, Boa, a model-turned-influencer with over 50,000 followers who started her business after becoming an influencer online, said that she not only fills her account with photos of her clothes, but also of her enjoying nights out, cooking, or getting manicures. She said she even posts long diary entries outlining her daily frustrations. The variety of content, according to Boa, is needed to cater to her followers who are not necessarily her clients, but who have been following her account from the time that she first gained popularity as a lifestyle influencer. Boa explains that she did not want to appear to be solely committed to making money but also maintain an image of a relatable lifestyle influencer who also 'happens to be' a businesswoman.

Entrepreneurs such as Kang and Boa aim to achieve the 'halo effect' for the reputations of their businesses by translating attributes unrelated to their products, such as 'rewarding personal relationships' or a respectable image (Abidin and Thompson, 2012: 470), into cultural and economic capital. In her study of Singaporean blogshops, Abidin emphasises the importance of creating an image of 'commonness' among blogshop owners to 'maintain readers' identification' by mentioning the mundane, such as 'complaining about daily household chores', and the 'hard work required to keep up their businesses' (Abidin and Thompson, 2012: 472). By putting their lives on display through the blurring of 'public and private lives' (Duffy and Pruchniewska,

2017: 853), these women engage in 'persona intimacy', a marketing tactic that cultivates 'emotional attachment not to the products per se but to the online personas' of the sellers (Abidin and Thompson, 2012: 472). Rather than 'product intimacy', a tactic traditionally used to 'build emotional attachments between customers and brands' through generating 'intensive love and long-lasting respect' from customers, building 'persona intimacy' has become more effective in achieving economic success among small-scale social enterprises that compete in an environment where many online sellers use similar suppliers for manufacturing products 'generally similar in design and style, and often even identical' (Abidin and Thompson, 2012: 469).

Alone in the cold: Protecting oneself from over-exposure

The need to constantly put one's life on display, however, and the fact that small enterprises lack the legal provisions of a traditional corporation also means that women entrepreneurs are more exposed to abusive situations that they have to deal with on their own, such as responding to hurtful comments or unethical customers. One Instagram shopping mall owner said that the ability to guard herself from being roped into situations that would require emotional labour is as important as maintaining an attractive image. She claimed that, as the only employee of her business, she felt unprotected at work. This is especially the case when the companies that develop the platforms are inherently 'on the customer's side', and offer no protections for the women entrepreneurs to effectively deal with difficult or unethical customers. She says that she avoids possible abuse from clients that may occur once they realise she runs a one-person business by telling them that her business is run by four to five people.

> If a customer calls me, I pretend that this business has a separate delivery team, customer service employee, manager, CEO, and model. I play all these roles myself. It is good to do this because it gives me protection and the customer does not look down on me. There is a difference between saying 'Sorry, I made a mistake in the delivery', and 'We apologise. We have a new employee on our delivery team so please excuse the mistake.' If I outright say, 'I will reimburse you', customers will ask for outrageous reimbursement. But if I say, 'I will try to get the CEO to approve of my decision to reimburse you', then the customers become more patient [...].

> If a complaint comes in, I fabricate a story about how I, a customer service representative, informed the CEO of the issue, and how the CEO became very upset and asked me to unconditionally offer a refund. I also say that the CEO chastised the entire staff. The customer may not show it, but I know they feel much better […] If customers know that there is a CEO overlooking everything, they are more cautious about how they treat me on the phone. […] If I go even further and say, 'I will secretly add some free socks to your order without the CEO knowing, as a way of apologising for our mistake', the customer becomes super happy. It's different from saying, 'I am the CEO and I will just throw in a pair of socks'. I know a lot about what makes the customer tick, happy, jealous, because I was once the customer that shopped a lot online. My blood pressure goes up for sure when I receive complaints, but I always remind myself, 'What would I have felt if I were in that person's shoes?'

On Instagram, it is easy to see female entrepreneurs having emotional outbursts, ranging from the apologetic to outright anger, in response to verbal abuse they receive from clients. In most cases, business owners, who are also the sole workers, must be the ones who directly apologise to clients if a shipment is delayed, argue with customers who return products in bad condition, or face abusive messages from clients who accuse the CEO of sharing personal hardship stories as 'sob stories'.[3] The lack of legal or financial resources to fight back is a good example of how the freedom that is promised through the ability to work for oneself can undermine other structural protections usually offered at corporate office jobs due to the lack of institutional provisions platform companies provide to e-commerce entrepreneurs.

Maintaining visibility: Bodily labour

As an image-based platform, Instagram appeals to women working in diverse industries, hosting images of goods and services offered by fitness instructors, fashion designers, painters, chefs, florists, photographers, and more. Yet, a simple Google image search of the words 'Shopping mall' (쇼핑몰), 'CEO' and 'Instagram' in Korean yields pages of uniform images of young, thin women in tight or revealing clothing, posing in suggestive ways at photo studios, upscale restaurants, and even tropical vacation spots. Though the models sell different products in different shops, they all share a similar physique: elongated, thin legs, big eyes, pale face, and hourglass waist, with disproportionately large bust and buttocks.[4] These self-chosen images are reflective of another form of immaterial work in which women entrepreneurs engage to stay afloat in the market: The visual labour to remain reputable online that ultimately ensures that 'heteronormative standards of beauty and emulation are reaffirmed rather than challenged' (Duffy, 2015: 454).

The emphasis on presenting a scantily clad, Barbie-esque body is also prevalent among women in the fitness industry such as sportswear designers, 'dieters' (self-proclaimed nutritionists), or fitness instructors. Many photos in the accounts used for this analysis are selfies taken in front of a mirror in which the model, clad in a tight-fitting sports bra and short workout shorts, would hide her face behind the camera but fully display her cleavage, bare abdomen, or bare thighs. Many workout videos, instead of being filmed from the front, are filmed from behind the woman, in some cases accentuating her butt by placing the camera on the floor, echoing the voyeuristic conventions of an upskirt shot taken in secret.[5] Even women in unrelated industries, such as florists, intersperse swimsuit and lingerie selfies among images of their products. This visibility tactic attracts a high number of followers in a short period, and these numbers translate into currency on social platforms as a high number of followers allows women to gather an audience to which they can then sell products. A higher number of followers also makes the account more attractive in the eyes of corporations that place advertisements on these business accounts as a way to organically advertise their products to a vast, loyal, and personal pre-existing audience.

The figure of the fashionable, sexy, successful woman CEO symbolises modern womanhood defined by 'success, attainment, enjoyment, entitlement, social mobility and participation' (McRobbie, 2008: 57) and enabled by a larger 'discourse of popular postfeminism' that proclaims the death of feminist struggles and encourages a 'raunch culture' through which young women idolise Playboy Bunnies and Victoria's Secret models as symbols of sexual liberation (Levy, 2006: 5). These marketing tactics, however, ultimately render many shopping malls spaces where female sexuality exists 'in relation to heterosexuality' rather than as an 'autonomous or independent sexual identity' (Gill, 2008: 51). The paradox of women attempting to remain sexually attractive under the guise of economic success has been dubbed the 'post-feminist masquerade', a 'highly styled disguise of womanliness' that adopts the 'too-short skirt' and the 'too-high heels' as a form of empowerment, but in reality is a form of women's anxiety over the possibility of completely losing male desire (McRobbie, 2008: 67).

The dominant aesthetic represented in these images suggest that women need to engage in 'a level of disciplining the body or physical maintenance', a type of 'aesthetic labour' or 'visibility labour' used to attract 'prospective employers, clients, the press or followers and fans' (Abidin, 2016: 90). Traditional tactics in the mass media such as the sexualisation of the female body in products such as 'alcohol, cologne, and cars' advertised to male audiences are now being used among female entrepreneurs in a 'hyper-feminized homo-social domain' of online social commerce (Abidin and Thompson, 2012: 474). Despite the lack of male presence or authority in these virtual spaces of work, women have already imbibed an 'internalised male gaze' by engaging in the sexualised display of their bodies and through the surveillance of each other's bodies (Abidin and Thompson, 2012: 474) to gain monetary profit. Not only does such aesthetic labour create an over-representation of young, thin, beautiful women who conform to mainstream aesthetic standards, it also perpetuates a cycle of women competing and labouring through bodily self-surveillance (Duffy and Hund, 2015: 9), disguised as a 'matter of personal choice' (McRobbie, 2008: 67).

Feminised roles

CEO, entrepreneur, mother: Redefining professional and domestic success

In Korea, a new group of women has taken centre stage online – the stay-at-home mothers with young children who recently left the job market to become full-time mothers, but have re-entered the workforce as owners of online businesses. The term 'mompreneur' is commonly used to describe these mothers who are seen to have successfully reconciled the duality of the 'productive' economic work of a career woman, and the 'reproductive' domestic duties of motherhood (Luckman, 2015). Mainstream media outlets in Korea have praised mompreneurship as an opportunity for young mothers to 'rid themselves of the label' of 'career-foreshortened woman' by setting up shops on Instagram. Newspapers feature the stories of stay-at-home mothers who became CEOs of their own companies thanks to Instagram lowering entry barriers for entrepreneurship, and to the availability of know-how on becoming a successful female online entrepreneur on Instagram, as mompreneurs themselves have said (Lee, 2017b).

A closer look at a sample of the Instagram accounts of mompreneurs, however, suggests that online micro-enterprise, rather than allowing mothers to 'have it all', may pressure mothers into 'doing it all'. Mothers sampled in the study create paradoxical façades to maintain favourability among their online audiences, including by attempting to appear as alpha female businesswomen while simultaneously reassuring audiences that they still prioritise their duties as mothers, and making sure they maintain the physical sexual attractiveness of the single, childless woman. Among the accounts selected for the analysis, this pressure is specifically manifested through images that suggest the omnipresence of children at the mother's workplace and through public denunciations of postpartum bodies that encourage women to reclaim the slim, sexy, prepartum physiques.

'Model' motherhood

@applekim2, @smitruti1010, and @luv__ribbon are prolific Korean mompreneurs on Instagram, each engaged in businesses of her own. @applekim2 advertises herself as a 'healthstagrammer', a 'lovely stay-at-home mom of two daughters + designer + working mom' with over 150,000 followers. She is a fitness and health enthusiast, and many pictures capture her workouts at the gym, meal plans, and post-childbirth weight-loss photos. @smitruti1010 is also a mother of two and a professional 'dieter' who describes herself as 'a mother who raises two sons and exercises with the youngest daughter in her arms'. With over 360,000 followers, she uploads mostly instructional Pilates videos and before-and-after postpartum weight loss photos of female clients who follow her weight loss programme. @luv_ribbon is an entrepreneur with over 410,000 followers, a mother of two, who specialises in retailing women's clothing and accessories.

Both @applekim2 and @smitruti1010 often feature children in their daily posts. Children are often seen running around their mothers during sportswear and fitness photoshoots, clinging to them during exercise, having fun on their mothers' exercise equipment, or on their mothers' backs while they are doing squats. Fashion retailers such as @luv__ribbon frequently feature her children following her to the wholesale market where she obtains her products.[6]

The rhetoric of motherhood is also displayed through self-shaming and public penance, as shown by @luv__ribbon, who blames her lack of oversight when her kids fall ill while she is at work, labelled '#sickstagrams':[7]

> I really don't want to write a sickstagram, but yesterday my son ate a fish cake daddy brought from Busan, and was sick all night, vomiting up all the medicine I gave him. Only when I gave him a heating pad for his stomach did his migraines disappear and he begin to eat again [...] I am sorry. I will be a better mother that serves only healthy meals and encourages exercise in the future.

Beneath the images of the ambitious mother is the reality of freelance work that defines mompreneurship, where the freedom to integrate work and passions comes at the cost of the impossibilities of separating childrearing from career building, as work and life have become fully intertwined, 24 hours a day. In these women's accounts, the father figure usually appears in photos taken outside the home, during family vacations or dinners at restaurants, and, in some cases, is absent from the account altogether. The reconciliation of business and domestic work, with its lack of corporate protections, structure, or benefits, may entail longer work hours and an even more intensive form of mothering that requires working women to spend entire workdays with their children (Taylor, 2015: 177). These images of increasing numbers of women playing the dual role of breadwinner and mother online exemplify the twenty-first century version of the 'housewife trap' (Taylor, 2015: 175), a term coined by feminist Betty Friedan in the 1960s to refer to a 'mystique of feminine fulfilment' that ultimately pushes women back into the 'home and domesticity' despite their 'hard-won fights for the right to escape from home' in the mid-twentieth century (Taylor, 2015: 183).

Hot on the market

In addition to displaying devotion to motherhood, mompreneurs on Instagram seem to be under additional pressure to maintain a youthful body and image. A common theme running through the blurbs of 'healthstagrammer' mothers' accounts treats postpartum weight loss as a goal for which mothers should strive. One mother in the sample lures followers with her weight loss story (18 kg lost after giving birth) and manages a business account that features weight-loss related foods and exercise equipment. Mompreneurs such as @smitruti1010 treat their weight loss as marketable assets online, as well as a source of personal pride in being both mothers who own businesses and wives who still possess sexually desirable bodies. Losing weight after giving birth to a child is usually displayed through before-and-after photos or videos.[8] In these photos, the 'before' photo usually features a woman before losing weight, looking sad, or expressing unhappiness through crying face emojis covering their faces. The 'after' photos often feature women in sexualised poses, with women sensually lifting their shirts to reveal muscular, slim midriffs or wearing tight shorts to display thin, waxed legs. As one mompreneur who works for Herbalife writes under her before-and-after photo:

> After I had a child, I felt very light and told my husband that I had lost all the weight, but he took a photo of me that made me face reality. It was a kick to my stomach, that I actually looked like this! No wonder people kept asking me if I was pregnant [...] I was depressed and kept asking myself whether I could continue like this when I found #Herbalife and lost #18kg and even made #abs. Doing it alone is hard but working together makes it easy. Ping me at #katalkhikanie for questions on my products!

In the comment threads below these posts, female followers post remarks such as, 'How do I get your legs?' or 'You look like the youngest aunt, never the mother!' or even, 'It is good to see you back (from being fat)!' Such interactions seem to have the effect of encouraging the entrepreneurs to present themselves in more objectified, sexualised ways. Most importantly, they reveal how female e-commerce on social platforms capitalises on the deflected male gaze enacted by women in the absence of men. These forms of oppression drive women to feel the need to be acknowledged online by identifying themselves with male norms of female beauty through a sense of self-shaming produced by social pressure – in this case, fuelled by women themselves – that demand that married women look like a pre-childbirth version of themselves. Such performances following the traditional gendered temporalities imposed onto a woman's lifetime are presented in commercial ways to assure audiences that they are being followed as a matter of personal choice, especially among the female-saturated audiences of Instagram, where 'patriarchy and hegemonic masculinities' have seemingly disappeared, but in reality have been overtaken by new cultural norms sustained by 'constant self-judgment and self-beratement' (McRobbie, 2008: 68).

If in the past, 'it was the home that was the ideal focus for women's labour and attention from which their "worth" was judged', today, it is her body that should 'be suitably toned, conditioned, waxed, moisturised, scented, and attired' (Gill, 2008: 42). In addition to the duties related to 'caring or nurturing or motherhood', the 'possession of a "sexy body"' has taken the spotlight as a woman's 'key source of identity' (Gill, 2008: 42) inside and outside the home. In the post-feminist era where women are seemingly given more freedom to rearrange the 'cycle of life-events' including taking control of their careers, marriage, and fertility, the type of social e-commerce activities Korean women engage in are in effect reinforcing traditional temporalities governing a woman's life. These perceptions are defined by the need for women to appear sexually attractive during their 'prime' for reproduction and marriage, and to become nurturing figures towards their husbands and children thereafter.

Ultimately, the 'explosion of micro-entrepreneurial home-based design craft labour' among young mothers represents a new façade of contemporary capitalism defined by 'new kinds of economic management' including new forms of domesticised labour that push women 'back into traditional roles within the home' (Luckman, 2015: 147). Promising online entrepreneurship as an 'answer to secure employment' may not be too different from waving a carrot in front of 'exhausted, over-committed' mothers (Luckman, 2015: 155) who cannot take advantage of maternity leave, find affordable daycare, or build transferable job skills (Luckman, 2015: 156). The 'control and choice often enabled by new technologies' that ostensibly allow the 'reconciliation of paid work and family responsibilities' may not always lead to personal and career fulfilment, but do illustrate the process of a 'post-Fordist precarity and institutionalized individualization' that takes the risk and responsibility for welfare away from society and places it in the hands of individuals, especially women (Luckman, 2015: 153).

Conclusion

Digital technology is in essence a 'sociotechnical product' that combines 'people, organizations, cultural meanings, and knowledge', and serves as a major 'source and consequence of gender relations' (Wajcman, 2010: 149). The analysis of the aforementioned Instagram accounts of Korean women entrepreneurs illustrates the dynamics of gender relations in the digital media economy, which no longer operate through a rhetoric of direct domination, but function more through self-driven bodily 'discipline and regulation' (Gill, 2008: 53). The very design of platforms such as Facebook, Instagram, and Twitter drive users to produce emotional responses such as giving likes, sharing posts, or expressing opinions through comments. In this environment, women are encouraged to generate profit by utilising marketing tactics reliant on gendered performances such as maintaining a caring, relatable reputation for online audiences or maintaining sexual attractiveness, factors that make work on digital platforms for women an extension of caretaking and aesthetic labour offline.

Immaterial, affective labour is not merely a form of 'creative autonomy' that may eventually bring future returns, but a form of work that is part of the market's rationalisation of the 'new forms of governance that shift risks from central organizations onto individuals' (Duffy, 2015: 453). Such forms of impalpable labour demand special scrutiny as they have the potential to further disseminate existing gender biases that deem women as 'more accepting of unpaid, unjust situations' (Duffy, 2015: 453). Rather than merely questioning 'how to promote women's participation at the workspace', dialogue is needed on 'how workplaces, organisational hierarchies and imagined ideas (have) become gendered' through the introduction of new technologies and work (Proctor-Thomson, 2017: 139). The first step toward this goal can be taken by advancing the focus of policy-making and academic discussion from the 'macro-analysis of women and employment' (McRobbie, 2011: 75) towards more nuanced conversations of the new forms of gendered practices and positions women must undertake when engaging in 'new forms of precarious work' enabled by digital technologies (McRobbie, 2011: 75).

To enable positive change, we must first question what it means when increasing numbers of young women flock to the net to become 'small-scale cultural or creative entrepreneurs' (McRobbie, 2011: 76). It should be acknowledged that structural issues such as internal gender bias, compulsory long work hours, and the lack of benefits are not merely obstacles that can be overcome 'by sheer grit and determination' (McRobbie, 2011: 76). The problem lies not within the industry or the small business itself, but in the ways the current neoliberal economy deploys self-employment and flexible labour as vehicles to privatise and personalise the responsibilities of the welfare state, such as 'retirement, caring responsibilities, unemployment and under-earning' (Taylor, 2015: 185), all under a pretence of a trend valorising social e-commerce as cool, meritocratic, and flexible.

The accounts studied show how work on digital media platforms is affected by a 'feminisation of work', a phenomenon that increasingly requires labourers to engage in 'emotional work and affective labour' historically assigned to women (Taylor, 2015: 182). In this environment, women continue to be given the responsibility for dealing with relationship-building and product circulation instead of roles in technology development and product production. A re-imagination of current technologies would include social encouragement from a young age for Korean women to pursue education in science and technology, subjects still deemed 'masculine' in Korean society. These beliefs have contributed to women making up a national average of only 17 per cent of students and 5 per cent of full-time professors in engineering departments at universities (Oh, 2016). Such low number of women in science 'contributes to maintaining a male-dominated culture of production' that places the highest value of technological skills of 'developers, engineers, venture capitalists, or entrepreneurs' above the 'marketing, public relations, project management, event planning, graphic design, or community management' jobs considered 'lower-status jobs' and mostly performed by women (Marwick, 2013; See also introduction to Part I about gender inequality in Korean society).

Additionally, treating workers as independent entrepreneurs rather than employees or encouraging starting a business over offering maternity leave or affordable childcare are ultimately acts that place the burden of social welfare in the hands of the individual. Sustainable change would involve policy-level reform to accommodate women returning to the workforce after giving birth as well as providing more protections for female entrepreneurs not only seeking support to set up their businesses on existing platforms owned by global corporations, but also striving to create their own digital spaces and tools to develop transferable skills outside of abstract, aspirational, and affective labour. Only by remaining cognisant of gendered practices that still pervade digital environments of work will there be opportunities for society to question and challenge new, technologically enabled capitalistic structures and move towards building a truly meritocratic, egalitarian work environment for all.

Notes

[1] 'Career Foreshortened Woman' 경력 단절녀 is a Korean neologism referring to women whose professional careers have been 'cut short' due to marriage and childbirth and are now unemployed full-time mothers.

[2] A 2-km strip of land located directly north of the Han River in Seoul, Dongdaemun Wholesale Market has served as the centre of textile manufacturing as well as fashion retail for the past century. Adjacent to a plethora of department stores, Dongdaemun Wholesale Market opens around midnight to accommodate the nearby small-scale retailers and online business owners who come to buy clothing and accessories to retail. An area first developed by Korean merchants in the early 1900s to unify against the growing influence of Japanese merchants in Jongno in central Seoul, Dongdaemun became the centre of textile trade due to the increased import and consumption of Japanese cotton. After the Korean War (1950–53), as foreign aid flowed into the country in the form of second-

hand clothing, Dongdaemun established itself as Korea's central wholesale textile market. Most of the merchants were defectors from the North, so stores were named to reflect their sentiment after the war, such as Pyeongwha (Peace) Market, which rose to become one of the main textile manufacturers of the 1960s (Shin, 2014). In the 1970s, Dongdaemun accounted for 70 per cent for the country's clothing manufacturing industry and became a full-time wholesale night market (10 pm–6 am) in the 1980s, after the curfew originally imposed by American military in 1948 was lifted. In the 1990s and 2000s, with the construction of large-scale retailer malls, Dongdaemun became the national centre of fast fashion and design (Kim, 2015). Today, Dongdaemun Wholesale Night Market has taken on a new role in the digital economy as the central supplier in the female-dominated retail social e-commerce ecosystem. The market is praised for having the best structure to accommodate the e-commerce industry as it offers a fast circulation of products, serves as the first entry point of new fast fashion trends, and forms a self-sufficient ecosystem comprising manufacturers and retailers of various textiles, leathers, and accessories in one place (Jeon, 2005). All interviewees in this study working as e-commerce entrepreneurs cited Dongdaemun as the primary supplier of online fashion-related businesses.

[3] @_yyesol, a prominent female shopping mall owner with over 106,000 followers, uploads pictures of texts she receives from clients: to an angry customer's text message complaining about a late shipment, the CEO apologises profusely on her public profile https://www.instagram.com/p/BS0w2rPjIzY/; a message from a follower criticising the CEO for using her personal stories about a family tragedy as a marketing ploy, https://www.instagram.com/p/BS9pyEPjUZb/; the 'personal story' that the follower criticised, @yyesol's personal post about her father's burned hand from a recent house fire, and the CEO responding in disbelief to being accused of using her family as a marketing tactic https://www.instagram.com/p/BS7LkDpDiqV/.

[4] Accounts of prominent female entrepreneurs focused on women's clothing, fitness, and diet products: https://www.instagram.com/__leeheeeun__/, https://www.instagram.com/ballet_jy/.

[5] https://www.instagram.com/p/B0c1zm7hS0i/, https://www.instagram.com/p/BJnY-sKh2qu/, https://www.instagram.com/p/Bp1RkbRgb0Y/.

[6] https://www.instagram.com/p/BxQ4l6CDeG1/, https://www.instagram.com/p/Binw9M6HFeg/, https://www.instagram.com/p/BgIgpyZAa3d/, https://www.instagram.com/p/BUs4ExKAb7j/, https://www.instagram.com/p/BnnuxSOg-zR/.

[7] https://www.instagram.com/p/BS-r_R1lIa6/.

[8] https://www.instagram.com/p/BhqeOEJgj3I/, https://www.instagram.com/p/BUV29Sug6l0/, https://www.instagram.com/p/BxbdDjXBR38/, https://www.instagram.com/p/BwGSohJDpvc/.

References

Abidin, C. (2016) 'Visibility labour: Engaging with influencers' fashion brands and #OOTD advertorial campaigns on Instagram', *Media International Australia*, 161(1): 86–100.

Abidin, C. and Thompson, E. (2012) 'Buymylife.com: Cyber-femininities and commercial intimacy in blogshops', *Women's Studies International Forum*, 35(6): 467–77.

Arcy, J. (2016) 'Emotion work: Considering gender in digital labour', *Feminist Media Studies*, 16(2): 365–68.

Cho, S. (2016) '이화여대-페이스북코리아, 여성 사업가 교육지원 협약 체결' [Ewha Women's University: Facebook Korea make agreement to support female entrepreneurship education], *Maeil Business Newspaper*, 13 June, Available from: http://news.mk.co.kr/newsRead.php?no=422565andyear=2016 [Accessed 22 March 2020].

Choi, S. (2016) '취업난과 경력단절에 초기비용 적게드는 쇼핑몰창업으로 몰려' [People flocking to open online shopping malls that require little initial capital amid low employment rate and career-foreshortening], *Maeil Business News Korea*, 27 November, Available from: http://news.mk.co.kr/newsRead.php?year=2016andno=822811 [Accessed 22 March 2020].

Cohen, D. (2017) '25 million businesses are now using Instagram', *AdWeek*, 30 November, Available from: http://www.adweek.com/digital/25-million-businesses-profiles-instagram [Accessed 22 March 2020].

Conor, B., Gill, R., and Taylor, S. (2015) 'Gender and creative labour', *The Sociological Review*, 63(1): 1–22.

Draudt, D. (2016) 'The struggles of South Korea's working women', *The Diplomat*, 26 August, Available from: https://thediplomat.com/2016/08/the-struggles-of-south-koreas-working-women/ [Accessed 22 March 2020].

Duffy, B.E. (2015) 'The romance of work: Gender and aspirational labour in the digital culture industries', *International Journal of Cultural Studies*, 19(4): 441–57.

Duffy, B.E. and Hund, E. (2015) '"Having it all" on social media: Entrepreneurial femininity and self-branding among fashion bloggers', *Social Media and Society*, 1(2): 1–11.

Duffy, B.E. and Pruchniewska, U. (2017) 'Gender and self-enterprise in the social media age: a digital double bind', *Information, Communication and Society*, 20(6): 843–59.

Elkins, K. (2015) 'From pumping gas to a $6 billion fortune – The impressive rags-to-riches story of Forever 21's husband-and-wife cofounders', *Business Insider*, 12 May, Available from: https://www.businessinsider.com/rags-to-riches-story-of-forever-21-cofounders-2015-5?r=USandIR=T [Accessed 22 March 2020].

Finn, G. (2015) 'Instagram has arrived as a haven for ad dollars, thanks to global ad rollout and new ad options', *Marketing Land*, 9 September, Available from: https://marketingland.com/instagram-has-arrived-as-a-haven-for-ad-dollars-thanks-toglobal-ad-roll-out-new-ad-options-141909 [Accessed 22 March 2020].

Gill, R. (2008) 'Empowerment/sexism: Figuring female sexual agency in contemporary advertising', *Feminism and Psychology*, 18(1): 35–60.

Go, H. (2018) '인스타그램만 나홀로 성장…국내 이용률 15%P 급증' [Instagram shows solitary growth…15% increase of domestic users], *Yonhap*, 15 March, Available from: https://www.yna.co.kr/view/AKR20180315064300017 [Accessed 22 March 2020].

Han, M. and Kim, Y. (2016) 'Can social networking sites be e-commerce platforms?', *Pan-Pacific Journal of Business Research*, 7(1): 24–39.

Hardt, M. and Negri, A. (2001) *Empire*, Cambridge, MA: Harvard University Press.

Hirsch, P. M. (1972) 'Processing fads and fashions: An organization-set analysis of cultural industry systems', *American Journal of Sociology*, 77(4): 639–59.

Hong, J. (2018) '네이버 점유율 71.5%…이용자 23%, 다음·구글로 바꿀 것' [Naver claims 71.5%…23% of users shifting to Daum, Google], *Yonhap*, 15 September, Available from: https://www.yna.co.kr/view/AKR20180919182000017 [Accessed 22 March 2020].

Jeon, C. (2005) '동대문시장 인터넷 쇼핑몰 개점 러시 – 인터넷 쇼핑몰로 활로 찾아…1천여개 쇼핑몰 운영' [Dongdaemun rushes to open shops online – Internet shopping malls take off…over 1,000 shops managed], *Ohmynews*, 25 August, Available from: http://www.ohmynews.com/NWS_Web/View/at_pg.aspx?CNTN_CD=A0000276554 [Accessed 22 March 2020].

Kim, D.G. (2016) '서울시도 '#그녀의 비즈니스를 응원합니다' [Seoul city also participates in #Shemeansbusiness], *Yonhap*, 24 October, Available from: https://www.yna.co.kr/view/AKR20161023051700004 [Accessed 22 March 2020].

Kim, G. (2018) '쇼핑몰 창업 – (8) 동대문 도매시장 사입 후 오픈마켓 판매' [Shopping mall entrepreneurship – (8) Selling on an open Market after buying from Dongdaemun Wholesale Market], *Naver Post*, 31 January, Available from: http://m.post.naver.com/viewer/postView.nhn?volumeNo=12639922andmemberNo=37459126 [Accessed 22 March 2020].

Kim, S. (2015) '[그때 그곳 지금은]⑦ "의류 메카" 동대문… 쇼핑몰·도매시장 명암 뚜렷' [Places then and now ⑦ 'Clothing mecca' Dongdaemun … Shopping mall and wholesale market have clearer picture], *Chosun Biz*, 16 November, Available from: http://biz. chosun.com/site/data/html_dir/2015/11/15/2015111500707.html #csidx5fbd9a60f7f3ec08dcfd0133e2ababe [Accessed 22 March 2020].

Kim, T. (2017) '네이버쇼핑 빠른 성장…2위 G마켓과 격차 좁혀' [Naver shopping shows rapid growth … closes gap with market no.2 G-market], *Yonhap*, 15 October, Available from: https://www.yna. co.kr/view/AKR20171013193100033 [Accessed 22 March 2020].

Kwon, S. (2017) '시청자 울렸던 '17살 얼짱' 쇼핑몰 사장님의 근황' [What the 17-year-old Ulzzang shopping mall CEO who melted our hearts is up to these days'], *Insight*, 20 May, Available from: http://www.insight.co.kr/newsRead.php?ArtNo=106334 [Accessed 22 March 2020].

Lagarde, C. (2017) 'Together, Korea's women and economy can soar', International Monetary Fund, 17 September, Available from: http:// www.imf.org/en/News/Articles/2017/09/05/sp090617-together-korea-s-women-and-economy-can-soar [Accessed 22 March 2020].

Lee, J. (2017a) '가장 많이 하는 창업 1위 '온라인 쇼핑몰' [Online shopping malls rank number one in new businesses opened], *NoCut News*, 16 January, Available from: http://www.nocutnews.co.kr/ news/4718673 [Accessed 22 March 2020].

Lee, J. (2017b) '인스타그램으로 '경단녀'에서 '멋진 창업맘' 됐어요' [Through Instagram, I became a 'cool mompreneur' instead of remaining a 'career foreshortened woman'], *Premium Chosun*, 16 May, Available from: http://premium.chosun.com/site/data/ htmldir/2017/05/16/2017051601156.html [Accessed 22 March 2020].

Levy, A. (2006) *Female Chauvinist Pigs: Women and the Rise of Ranch Culture*, New York: Free Press.

Luckman, S. (2015) 'Women's micro-entrepreneurial homeworking', *Australian Feminist Studies*, 30(84): 146–60.

Marwick, A. (2013) 'Silicon Valley isn't a meritocracy: And it's dangerous to hero-worship entrepreneurs', *Wired*, 25 November, Available from: https://www.wired.com/2013/11/silicon-valley-isnt-a-meritocracy-and-the-cult-of-the-entrepreneur-holds-people-back/ [Accessed 22 March 2020].

McRobbie, A. (2008) *The Aftermath of Feminism: Gender, Culture and Social Change*, Thousand Oaks, CA: Sage.

McRobbie, A. (2011) 'Reflections on feminism, immaterial labour and the post-Fordist regime', *New Formations*, 70: 60–76.

Min, J. and Jung, Y. (2017) '동대문 온라인몰 창업신화 쓴 스타일난다... 로레알·시세이도까지 '군침'' [Stylenanda, a Dongdaemun online mall miracle ... attracting the interest of L'Oreal and Shisheido], *Korea Economic Daily*, 4 December, Available from: http://plus.hankyung.com/apps/newsinside.view?aid=2017120487 021andcategory=andsns=y [Accessed 22 March 2020].

Negra, D. and Tasker, Y. (eds) (2007) *Interrogating Postfeminism: Gender and the Politics of Popular Culture*, Durham, NC: Duke University Press.

Oh, W. (2016) '한국 IT업계 남녀 임금 격차 살펴보니' [Looking at the gender wage gap in Korea's IT industry], *Bloter.net*, Available from: www.bloter.net/archives/247736 [Accessed 22 March 2020].

Proctor-Thomson, S. (2017) 'Feminist futures of culture work? Creativity, gender and difference in the digital media sector', in M. Banks, R. Gill and S. Taylor (eds) *Theorizing Cultural Work: Labour, Continuity and Change in the Cultural and Creative Industries*, London: Routledge, pp 137–48.

Sengupta, S., Perloth, N., and Wortham, J. (2012) 'Behind Instagram's success, networking the old way', *The New York Times*, 14 April, Available from: https://www.nytimes.com/2012/04/14/technology/ instagram-founders-were-helped-by-bay-area-connections.html [Accessed 22 March 2020].

Seoul Institute (2017) '서울시 청년들의 "취업'과 '창업" (서울인포그래픽스 제222호)' ['Employment' and 'entrepreneurship' of Seoul City's youth (Seoul Infographics – issue no. 222)], 13 February, Available from: https://www.si.re.kr/node/56888 [Accessed 2 January 2021].

Shin, Y. (2014) '동대문시장의 변화와 발전과정: 동대문 제5의 물결' [The changes and development of Dongdaemun market: The 5th Wave of Dongdaemun], *Dongdaemun Fashion Town*, 12 May, Available from: http://www.dft.co.kr/bbs/board.php?bo_ table=info5andwr_id=12andpage=2 [Accessed 22 March 2020].

Taylor, S. (2015) 'A new mystique? Working for yourself in the neoliberal economy', *The Sociological Review*, 63(1): 174–87.

Tokumitsu, M. (2014) 'Stop saying "do what you love, love what you do": It devalues actual work', *Slate Magazine*, January 16, Available from: https://slate.com/technology/2014/01/do-what-you-love- love-what-you-do-an-omnipresent-mantra-thats-bad-for-work-and- workers.html [Accessed 22 March 2020].

Wajcman, J. (2010) 'Feminist theories of technology', in S. Jasanoff, G.E. Markle, J.C. Peterson, and T. Pinch (eds) *Handbook of Science and Technology Studies*, Thousand Oaks, CA: Sage, pp 143–52.

Wilkie, R. (2011) *The Digital Condition: Class and Culture in the Information Network*, New York: Fordham University Press.

World Economic Forum (2017) 'The Global Gender Gap Report 2017', Available from: http://www3.weforum.org/docs/WEF_GGGR_2017.pdf [Accessed 2 January 2021].

2

'For Japan Only?' Crossing and Re-inscribing Boundaries in the Circulation of Adult Computer Games

Patrick W. Galbraith

Introduction

This chapter considers how and why 'Japan' comes to matter in the circulation of games in a globalised world. This may seem counterintuitive. As media scholar Mia Consalvo argues, computer/console games appear to be a 'global' and 'hybrid culture' (Consalvo, 2006: 117). However, Consalvo also highlights the substantial influence of Japanese hardware companies such as Nintendo and Sony and of Japanese franchises such as Super Mario Bros and Final Fantasy on the global gaming industry and culture. That is, Japan is already part of a global and hybrid gaming industry and culture; content produced in Japan is not typically labelled as 'foreign' or 'Japanese', which is remarkable in comparison to, for example, manga (comics from Japan) and anime (cartoons from Japan). Furthermore, Consalvo (2006: 120) draws attention to the fact that localisation and a softening of what is perceived as '*too* foreign' aids in the circulation of content produced in Japan (see also Allison, 2006). All of this is to suggest that perhaps 'Japan' does not matter as an organising category in the mainstream global gaming market, even though companies headquartered in Japan have helped shape that market and still deeply impact it.

This de-emphasising of Japan is not, however, common in discussions of adult computer games. On the contrary, many insist that these games, which range from simulated conversation to explicit sex with manga/anime-style characters, are somehow distinctively 'Japanese'. In her monograph on games and globalisation, Consalvo herself almost appears to agree, but then suggests that the putative 'Japaneseness' of these games might be better understood in terms of 'the contexts of their production' (Consalvo, 2016: 74). One must thus consider the confluence of computer games, adult content, and the growing desire for cartoon characters in manga/anime subculture in Japan since the 1980s (Galbraith, 2019). If manga, anime, and computer/console games are often mass media or culture, then they are also part of popular culture as what theorist Stuart Hall calls a 'terrain of [...] struggle' (Hall, 1998: 452). This struggle is an ongoing process wherein certain forms come to prominence and are promoted while others are policed and pushed aside. This process can lead to the formation of a 'national-popular culture', which should not be taken for granted, but rather seen 'as a battlefield' (Hall, 1998: 451). Likewise, subculture is not simply low or disposable culture – which is the sense in which the word is often used in Japanese discourse – but rather refers to 'noise' or 'interference in the orderly sequence' (Hebdige, 2005: 355), which can give rise to unexpected meanings and challenges. Manga/anime subculture, then, indicates disruptive engagements with and articulations of mass media and culture.

Sexual desire for manga/anime characters is one historic example of the noise of subculture, with fans harbouring such desire labelled '*otaku*' in the early 1980s. Adult computer games were developed by and for such fans, and the high concentration of companies producing and selling these games in and around the Akihabara area of Tokyo contributed to its transformation into an '*otaku* city' in the late 1990s and early 2000s. Conducting fieldwork in Akihabara since 2004, with a focus on adult computer games from 2014 to 2015, I have observed the area's subsequent transformation into a destination for international tourists and symbol of 'Japanese popular culture' and 'Cool Japan' (see Introduction about government policy). Even as Akihabara becomes a site to promote the manga, anime, and games that circulate globally and have drawn Japan fans around the world, the embedded subculture of adult manga, anime, and games inspire criticism and demands for stricter regulation. Pressure to abide by 'international standards' has been provocatively described by artists, activists, and even academics as 'a new form of cultural colonization' (Pelletier-Gagnon and Picard, 2015: 39).

Furthermore, adult computer games are increasingly circulating beyond brick-and-mortar stores in Akihabara and other areas associated with *otaku*. Even if adult computer games are not 'destined for global consumption' (Consalvo, 2006: 120), to borrow Consalvo's turn of phrase, they are crossing borders due to a combination of official and unofficial translation and distribution efforts online. As they reach audiences more familiar with types of games that are 'destined for global consumption', adult computer games are persistently labelled as 'Japanese'. This is a form of what media scholar Koichi Iwabuchi calls 'inter-nationalism', or the 'reworking and strengthening of the national in tandem with the intensification of cross-border media flows' (Iwabuchi, 2010: 89). To be precise, it is an inversion of the phenomenon of nation branding observed by Iwabuchi, wherein those in positions of authority nationalise content from the top down. His case study of the battery of policy and public diplomacy initiatives under the banner of 'Cool Japan' reveals attempts to associate desirable content with the nation and attract 'Japan fans'. This nationalisation of content and fandom, Iwabuchi argues, cuts off transnational connections, denies hybridity and smooths out internal differences and conflicts. In this way, anime, for example, is marketed to international consumers as 'Japanese'.

In contrast, the nationalisation of adult computer games occurs in reception and feedback, often from the bottom up and from multiple, dispersed sites in Japan and beyond. The nationalisation is part of a process that occurs on a terrain of struggle (Hall, 1998: 452). To put it another way, 'Japan' emerges in the noise generated by transmission and resistance to certain content (Novak, 2013: 10, 24). Even as meanings become unstable, statements that are taken for granted in discourse about Cool Japan are opened to interrogation; one can see how adult computer games come to matter in discussions of 'Japan', which is a contingent articulation in an ongoing process of crossing and re-inscribing borders. While Consalvo cautions against 'foolish and dangerous' attempts to determine '"essential" or "fundamental" national qualities that may be found in individual games' (Consalvo, 2006: 127), this is exactly what happens when adult computer games, even extreme examples, are taken by journalists, activists, and even academics in the Anglophone sphere to reveal the 'perversion' and 'social illness that's embedded in Japanese society' (Alexander, 2009).[1] As this content crosses borders, which gets some pundits crowing about Cool Japan, critics in other nations respond to Japan as an imagined source of danger and perversion (Hinton, 2014: 56, 65). This points us to, first, the re-inscription of national and regional borders through regulation and new regulatory regimes, and second, the push and pull of (inter-)nationalised content across borders.[2]

A brief note on adult computer games

Before the main discussion, a brief note on adult computer games is needed. In Japan, 'adaruto' (adult) or 'ero' (erotic) games are a niche market that is distinct from the mainstream of globally recognised franchises such as Super Mario Bros, Final Fantasy, and Pokémon (Mizuho Survey, 2014: 117–18). Labelled 'R-18', only players age 18 or older can purchase them. Adult games in Japan are usually played on personal home computers rather than on consoles or mobile devices; they are not networked or multi-player games like many examples of computer games in other parts of the world. Featuring graphics that are rarely computer-generated or realistic, they also do not allow players to freely roam in open environments. Instead, players read on-screen text and make choices in something approaching choose-your-own-adventure novels. Despite their seemingly rudimentary and outmoded design, newly released adult computer games cost around US$100 each. In recent years, adult computer games have had difficulty attracting new players, and the market has been shrinking for some time; today, most games sell only 1,000–2,000 copies (Kagami, 2010: 136; Yano, 2014; Sakakibara, 2016). On the one hand, even as mobile games and other casual, networked and multi-player forms are seeing massive growth, players are less willing to commit to long hours alone reading text on a computer screen; on the other hand, distribution has remained stubbornly material, with game software released on discs in bulky packages and loaded onto personal computers. While this commitment to the material was intended to ensure companies profited from sales and avoided running afoul of regulators, it also limited circulation and drove up prices. As the player base aged and shrank overall, it further drove up prices and discouraged newcomers not already committed to the format and its offerings.

As a niche, adult computer games appeal to dedicated players with specific demands. In terms of adult content, these specific demands include a focus on manga/anime-style characters, which can be involved in any imaginable form of sexual interaction. Given that Japan maintains a legal distinction between virtual and actual forms of pornography, and has legally limited the definition of obscenity to overly exposed genitals as opposed to acts, it is possible for adult computer games to depict sex involving underage characters, which would not be tolerated in other parts of the world (Galbraith, 2017). It is worth pointing out, however, that industry insiders estimate that adult computer games focusing on extreme sexual expression make up only 10 to 20 per cent of the market (Kagami, 2010: 128). These are also not the bestselling titles; those concentrate on interpersonal relationships, romance, and melodrama. Regardless, most include explicit sex scenes. Some game makers remove these scenes for later 'general' releases on home consoles, and the overall industry trend is away from hardcore content, but sex is still very much a part of adult computer gaming in Japan.

More than sex, adult computer games in Japan are fundamentally defined by their focus on interactions with characters, most prominently *bishōjo* or 'cute girls'. With interaction in mind, there are eight identifying features of adult computer games: (1) characters that the player character interacts with are seen from a first-person perspective and speak directly to the player character, who is for the most part unseen; (2) important characters will often have a recorded voice for lines of dialogue, but the player character does not; (3) characters appear as a series of still images, for example changing from one static facial expression to another as the interaction proceeds; (4) designed to be visually appealing, the characters have a distinctly cartoonish look that brings to mind manga and anime; (5) below characters on screen is scrolling text in a box, which tells a story that frames and motivates interactions; (6) the story takes at least several hours to complete; (7) along the way, the text presents the player with choices that can impact character interactions and the story; and (8) there are multiple possible endings. Over the course of the game, the player interacts with characters in ways ranging from casual conversation to coitus. Gender and media scholar Emily Taylor summarily describes adult computer games as 'interactive anime/manga with erotic content' (Taylor, 2007: 198). Interactive refers to making choices in textual branches of stories that can be quite long and involved, so it is perhaps not surprising that adult computer games are also sometimes called visual novels or novel games. For many overseas, however, they are simply '*hentai* games' (Bogost, 2011: 107) or 'pervert games from Japan' (Game Faqs, 2017).

The *RapeLay* controversy

Illusion's *RapeLay* (Reipurei, released in 2006) is, as the name suggests, an adult computer game about rape. Players molest women and force them into sexual activity on a crowded train and in station bathrooms.

In the end, the women, some high-school aged and dressed in school uniforms, seem to enjoy it. A niche game in a niche market, *RapeLay* did not rank among the top 50 most popular of the year and quickly disappeared from shelves in Japan (Getchuya, 2006). That was the end of it, until a pirated digital copy appeared without the consent of the producers from a reseller on Amazon.co.uk in 2009, sparking debate worldwide. It began with a report in the Irish newspaper the *Belfast Herald* on 12 February 2009, which featured comments from Keith Vaz, a politician in the British Parliament and strong advocate of game regulation, who vowed to pursue the issue (Fennelly, 2009). Vaz did bring *RapeLay* up with Siôn Simon, a member of the European Parliament, who suggested that pan-European steps against such content would be 'the building block to moving towards a global regulatory future' (Game Politics, 2009). In May 2009, the New York-based human rights group Equality Now launched an international protest campaign under the banner 'Japan: Rape-simulator games and the normalisation of sexual violence', which urged its 30,000 members in 160 countries to write letters of protest to Illusion and the Japanese government (Equality Now, 2009). 'These sort of games that normalise extreme sexual violence against women and girls have really no place in our communities', Taina Bien-Aime, executive director of Equality Now, told CNN. 'What we are calling for is that the Japanese government ban all games that promote and simulate sexual violence, sexual torture, stalking and rape against women and girls. And there are plenty of games like that' (Lah, 2010). In Japan, the commotion led to the Komeito Party, part of the ruling coalition with the Liberal Democratic Party, and the women's committee of the Liberal Democratic Party to hold meetings in May 2009 to discuss regulation of rape simulation games. One month later, the Japan Ethics Organization of Computer Software, an industry organisation for adult computer games, self-imposed a ban on the production and sale of rape games, as well as limiting the sale of adult computer games generally to Japan (Mainichi, 2009). This was the strongest possible action they could take, but it did not please everyone concerned.

Responses to the ban in Japan

In 2009, Nakasatomi Hiroshi, professor of law at Tokushima University and co-founder of the Anti-Pornography and Prostitution Research Group, published an article titled '*Reipurei mondai no kei'i to hō kaisei no kadai*' (The particulars of the *RapeLay* issue and the task of legal reform). Broadly speaking, Nakasatomi advanced three critiques of

RapeLay: One, the content is discriminatory against women; two, it constitutes a form of child pornography; and three, such games create a culture of sexual violence that puts women and children at risk. Nakasatomi considered the response of the Japan Ethics Organization of Computer Software insufficient for three reasons: One, the Japan Ethics Organization of Computer Software is a voluntary industry organisation without compulsory membership, so not all game producers are part of the group and in any case are not compelled to abide by its decisions; two, independent producers can use the Internet to distribute their works and circumvent official zoning; and three, the decision to ban *RapeLay* was reactionary and served to keep government regulators from imposing legal restrictions, which are necessary to close loopholes (Nakasatomi, 2013: 3–5, 7).[3] Nakasatomi's primary criticism is that adult computer games such as *RapeLay* are 'harmful to society' (Nakasatomi, 2013) and contribute to a culture of violent sexual oppression of women and children.

While Nakasatomi considered adult computer games to be a social problem in Japan, Kagami Hiroyuki argues that they are in fact a social good. Kagami has worked in the adult computer game industry for over 20 years, beginning in 1995 at Illusion, the production company behind *RapeLay*. Well-known and respected in his field, he is the author of *The Adult Computer Game Scenario Bible* (*Bishōjo gēmu shinario baiburu*, published in 2009) and has worked on around 100 adult computer games in various capacities. By Kagami's estimation, the *RapeLay* controversy demonstrates how public and private individuals are using the discourse of harm to legitimise a massive expansion of their authority over the lives – even the imaginary lives – of others. For his part, Kagami does not see the existence of adult computer games and similar content as a social problem in Japan, but rather considers Japan a model case of liberal sexual expression and low sexual crime:

> There is no other country where pornography is as developed as Japan. In the entire world, Japan is where eroticism developed most. Following from this, I think that this country has developed mechanisms to curb social unrest and stabilise society. As a porn novelist, I think that this should make us proud rather than ashamed. […] I surmise that the low rate of rape in Japan is due to Japan's rich culture of pornography. (Kagami, 2010: 182, 319)

Note how this celebration of adult computer games is every bit as nationalised as the critique of 'Japanese pornography'. Kagami wonders why the world does not consider the possibility that it has something to learn from Japan, but respects that it is not up to the adult computer game industry to impose anything on others.

That is why the circulation of adult computer games is controlled so strictly by the very industry that produces these games. Kagami highlights the '*bishōjo gēmu no rinri-teki na hairyo*' (ethical considerations of adult computer games), namely how the content is distributed in such a way as to mark it as separate from reality and not for everyone. For example, *RapeLay* displayed the following warnings on its packaging, in its manual, and during the loading of the game software:

> Please be aware that this game contains extreme expressions such as violence, cruelty and criminal acts. Molesting people on the train and rape are crimes when actually committed. These acts should absolutely not be mimicked.
>
> To stores: This game is intended for people age 18 and older. We request that as much as possible you display it in a section apart from general games or on a separate shelf where children and people uninterested in the content will not be exposed to it.
>
> The content of this software is to the last a constructed work of fiction (*sōsakubutsu*) and a game. Conducting the same acts depicted in the game in reality will result in punishment by law. The content of the game is a bit of theatre (*shibai*) and fiction, so absolutely do not be influenced by or mimic these acts.
>
> JAPAN SALES ONLY. (Kagami, 2010: 219–20)

As Kagami points out, the production and distribution of *RapeLay* was not a crime in Japan. The crime was someone unrelated to the company and its network of retailers selling the game online, outside the context of its ethical distribution and play. Apparently, the borders of this ethical distribution are not only the sections of stores zoned for 'adults', but also the nation of 'Japan'. That is, adult computer games are only appropriate for Japan. This was a discourse that I encountered often in the field.

'For Japan only'

Beginning a new leg of my ongoing fieldwork in Akihabara in 2014, one of my first tasks was to play adult computer games. I was told that they would not work on my Macintosh laptop, so I purchased a cheap personal computer from Yodobashi Camera in Akihabara. I also picked up some adult computer games, based on buzz I heard on the street and recommendations from the staff at speciality shops. That evening, after returning to my rented room in the western suburbs, I tried to load the game software onto my new computer, only to be met with compatibility errors. After trying to load every game I now owned multiple times, I finally gave up and resigned myself to asking for help the next day. Rather than go back to Akihabara, I stopped at a nearby Yodobashi Camera location, which served the suburban community. With my computer and receipt, one of the games in its package, and the user manuals in hand, I arrived at the help desk to find two young men helping an elderly woman set up her computer. Although I was in sweatpants and scowling due to my frustrated efforts the night before, the young man who turned to assist me was nothing but smiles – until he saw the software that I needed to load. With the elderly woman next to us, we tried to discretely discuss the problem, but he could not solve it and so called his colleagues over for advice. Soon there was a throng at the desk, as well as curious onlookers. As the minutes turned to an hour, I was blushing deeply. Finally, the prognosis. 'Japan only', the original young man says in English, before switching back to Japanese, 'That's the problem.' But I am in Japan. 'No, no, no. Japanese only.' But I am speaking Japanese. 'No, no, no. The computer's operating system must be set in Japanese.' With some wizardry, a change is made in the machine and suddenly the opening screen of the game pops up: two cartoon women with used condoms between smiling lips and semen spilling onto their breasts. My blush deepens, but thankfully this all took long enough that the elderly woman had long since departed.

The experience left an impression on me, but I got used to seeing the words 'Japan only' on the packages of adult computer games, in manuals and during the loading of software. The fact that there had been some sort of a compatibility problem with my computer and the software became a joke that I told gamers for a laugh, and they congratulated me on working out my issues (with the machine or myself, I was not sure). Hayase Yayoi, a voice actress working in the adult computer gaming industry, was among those who found the story funny and liked to spread it around. She told me so over a late lunch in Akihabara before we made our way to Charara!!, a monthly industry event.[4] We arrived a little early because Hayase wanted to introduce me around; she was not making an appearance that day, so she took me in the back elevator used by staff. While meeting people and exchanging business cards, I noticed that the event organisers seemed agitated by my presence. Identifying myself as a researcher from the US did not help. Later, Hayase explained that the organisers had decided that I was a journalist and that my reporting would be '*gaikoku muke*' (for foreigners). They did not want such coverage, because the Charara!! event is for producers to pitch games to core players who then make direct purchases, and because adult computer games are not for sale overseas, whatever I wrote would serve no purpose. If anything, it might cause trouble by encouraging people outside Japan to illegally distribute games online, and perhaps draw attention and negative publicity, as in the *RapeLay* controversy, which would hurt the industry.

Charara!! is a platform where adult computer game producers gather every month to show new work, generate buzz, and get support from fans. They show work in progress, sell samples or '*taikenban*' (testers) and invigorate the fan base by giving away prizes, staging signings with illustrators and hosting mini-lives with voice actresses and singers. Charara!! is small (about 50–100 men in the audience each month), but it is important in the sense that it builds relationships with an inner circle of fans who buy into the productions and support them. If the event organisers had known who I was, they would have refused to have me as a pre-registered guest, which is precisely what they did in a later email that reminded me that these games are '*Nihon kokunai hanbai nomi*' (for sale in Japan only). The Charara!! platform is not open access, but is rather limited, as is necessary for its continued existence. Violating the boundaries of the circle constitutes a threat. This was more so in my case, because bringing adult computer games outside the circle might confront others with content that troubles them and bring attention to the event, which would cause trouble for those inside the circle.

Despite being free and open to anyone, Charara!! is in fact a controlled and closed space, which allows for interactions between producers and players. It is not private, but it is also not public; a sort of private publicity or public privacy is key. Keeping the space open requires controlling its boundaries, which resonates with what anthropologist Chris Kelty calls 'recursive publics' that have 'a shared sense of concern for the technical and legal possibility of their own association' (Kelty, 2005: 192). As Kelty sees it, many groups are now concerned with their own 'conditions of possibility – and the modes of manipulating them technically and legally – on and off the Internet' (Kelty, 2005: 204), which is precisely what one sees at Charara!!. The conditions of possibility for the event are a closed circle with clear boundaries and controlled access. The event is 'for Japan only', specifically core players such as those who gather in Akihabara, and should not contribute to distribution beyond established boundaries. Responding to the threat of regulation from the outside, the industry regulates from the inside and insists on boundaries.

The Sekai Project

Not all circles are so closed or protective of boundaries. In the course of fieldwork, I met people such as Peter Payne of J-List, who lives in Japan has been selling adult computer games through his popular website since 1998, and Francesco Fondi, who founded *Play X*, the first adult computer game magazine outside Japan, in 1998 and went on to work in the industry in Japan for several years.[5] Both men underscore the global demand for and interest in adult computer games from Japan. I also met producers who knew their games were being pirated and distributed overseas, but who, rather than punishing fans, instead asked them for information about the limits of official distribution in their home countries and ways to make distribution legal. That is, they turned a blind eye to illegal distribution in exchange for market research and activated fans as advocates to spread the culture of adult computer games. This also reduced their own economic and legal risk, because producers can test the market without direct investment or company action that might invite legal responses. These efforts are opening new routes of circulation of adult computer games.

For example, Front Wing, a production company in Akihabara where I worked, decided to officially release its game *The Fruit of Grisaia* (*Gurizaia no kajitsu*) overseas. Fan translations of the game were already circulating through unofficial channels online, but Front Wing wanted to reconnect players to the company and establish revenue streams leading back to it. With an anime adaption of the game targeting general audiences set to air on Japanese television in 2014, *The Fruit of Grisaia* was an attractive property, and Front Wing took as a given that fans overseas would access the anime via illegal streaming sites and hence be curious about the original source material (including the excised sex scenes). That said, Front Wing understood that brick-and-mortar stores overseas simply would not stock such a computer game, and that mainstream sites such as Amazon would likely not take the risk of further controversy associated with 'pervert games from Japan'. Instead, Front Wing opted to go through the crowd-funding site Kickstarter to collect monetary support from fans in exchange for premiums at the time of release. To connect to this overseas base, the production company turned to the Sekai Project, founded in 2007 and headquartered in Los Angeles, which makes a business out of activating fan networks to translate, support and purchase adult computer games for digital download.[6] '*Sekai*' means 'world', which is fitting, seeing that their success made *The Fruit of Grisaia* available around the world.

Importantly, this was not global distribution planned from the start (Consalvo, 2006), but rather an after-the-fact attempt to make already existing content available in new markets through collaboration with fans. Behind the spread of adult computer games from Japan beyond its borders is what has been called 'progress against the law' (Leonard, 2005: 281), whereby fans desiring content draw it into new circulations despite a lack of official releases or even official channels for release. This is very much how the market for anime was established in North America, for example, with early adopters spreading the material through clubs and mailing lists, starting localisation companies and generally translating for others. As in the case of anime, this new content being drawn across borders can challenge dominant norms and values and lead to debates and emergent alternatives (Eng, 2012: 100–3; see also McLelland, 2005). In the case of the content itself being illegal, as in the case of some adult comics, cartoons and computer games from Japan crossing over into 'virtual child pornography', what anthropologist Rosemary Coombe calls 'juridical resolutions of meaning' (Coombe, 1998: 45) confront divergent understandings in groups, communities, and (sub)cultures connecting and communicating online. And there are signs of changing minds, from adult computer games (minus explicit content, but with links to easily add it after purchase) making the list of bestselling titles on mainstream digital distribution site Steam (Grayson, 2019) to indie games such as *Doki Doki Literature Club!* (developed by Team Salvato and released in 2017) not only garnering critical acclaim but also demonstrating a mastery of the genre by North American developers. Indeed, from its title to its character aesthetics, *Doki Doki Literature Club!* is an adult computer game produced in the style of Japan, but outside that country, in dialogue with producers and fans within and beyond its borders.

Bamboo dreams of the world

People say a lot of things about Bamboo (aka Takeuchi Hiroshi), but never that he does not dream big. A creative force at OVERDRIVE, an adult computer game company, Bamboo got his start working at Front Wing in 2000 and then produced *Green Green* in 2001. He believes that adult computer games have the power to impact and change lives, which gives him a sense of purpose and pride in his work. 'We are influenced by media', Bamboo tells me at his office in Asakusa, a historic temple and entertainment district in downtown Tokyo. 'Look at you. You wouldn't be here in Japan if it weren't for some media that influenced you.'[7] What really sets Bamboo apart from many of his peers in the industry is his dedication to expanding the adult computer game market overseas. With the premise that there are 'no borders' (he says this in English to underscore the point) on fandom, in 2008, Bamboo founded MangaGamer, an online store that takes as its mission bridging linguistic and physical boundaries and bringing adult computer game producers and players together.[8] A broad coalition of producers is now making their games legally available to players overseas through the MangaGamer website.

Although there are still questions about whether or not the possession of this content is legal in all the countries where it can be purchased and downloaded (Galbraith, 2017), Bamboo says that the site has never received a complaint. Instead, he emphasises feedback from those happy to finally get content that is too niche to make it into brick-and-mortar stores where they live. In its fifth year when we met, MangaGamer had 20,000 registered users, primarily from the United States. Bamboo not only does live streaming to communicate in English with adult computer game players in California, but also speaks at conventions such as the Los Angles Anime Expo (before a crowd of 800) and goes to dinner with fans to ask about their interests in games.[9] 'The people who like adult computer games are the same everywhere', he explains. 'From their appearance to their actions, they are the same. It doesn't matter where you are. When these fans are around, it's enough to make you think, "Is this Japan?"' Bamboo's imagined global community and culture of adult computer games stands in stark contrast to the discourse of 'Japan only', both in the sense of 'only in Japan, that crazy country' and 'for play and sale only in Japan'. Indeed, Bamboo seems almost to undo the nation itself, because anywhere that adult computer gamers come together feels like 'Japan'. On the one hand, the coming together of people in love with manga/anime-style cute girl characters creates a feeling of 'Japan'. On the other hand, the nation is something imagined in interactions and relations with manga/anime characters and one another. This is 'Japan' as game, or rather as an erotic game, one imagined and created collaboratively.

While others worry that the Tokyo Olympic Games will result in increased international scrutiny of manga/anime subculture and a crackdown on objectionable content, Bamboo instead sees an opportunity. In 2011, he produced *Go! Go! Nippon: My First Trip to Japan*, which uses the operating system and mechanics of an adult computer game (i.e., interacting with cute girls while reading text and making choices) to offer a virtual tour of Japan. Available for US$10 on Steam, *Go! Go! Nippon* contains both English and Japanese options and is intended to be a learning tool. By October 2014, the virtual journey through an imaginary Japan guided by cute girls had sold 30,000 copies. Bamboo dreams that sightseers might even use a mobile version of the game while travelling the country during the Olympics. Scholars have suggested that adult computer games can contribute to a sense of being 'tele-present in Japan' (Jones, 2005), and Bamboo takes it further by layering fiction onto reality. What 'Japan' means here is deeply ambiguous (Napier, 2007), but there can be no doubt that it powerfully attracts some and repulses others. For his part, even as he draws gamers to Japan in interactions and relations with manga/anime characters, Bamboo also goes beyond Japan to encounter them elsewhere. If Bamboo wonders in his overseas encounters 'Is this Japan?', then players might also ask the same question of the imagined and created world of *Go! Go! Nippon*. To seriously ask such a question is to shine a light onto the ongoing process of crossing and re-inscribing borders in the global circulation of adult computer games today.

Conclusion

I had long admired Watanabe Akio as an artist. Watching anime series such as *The SoulTaker* (2001), *Nurse Witch Komugi* (2002) and the Monogatari series (from 2009), I fell in love with his cute characters and colourful aesthetics. Perhaps I should have associated him with adult computer games earlier, for example when watching the anime series *The World God Only Knows* (from 2010), which features characters designed by Watanabe in a story about an adult computer gamer, but it was only in 2014, while I was working at Front Wing, that things clicked. The staff at Front Wing were excited that their game *The Fruit of Grisaia* had been adapted into an anime series and would air on television later in the year. The character designer for both the game and the anime was none other than Watanabe Akio, who I was told had worked in the industry since the 1990s on games such as *Angels in the Court* (*Kōto no naka no tenshitachi*), *Popotan* (2002) and *To Heart 2* (2004). Watanabe spoke with me over noodles near Ueno Park:

I love games far more than anime. I started working in the anime industry in 1986, and in 1995 I did my first character design for a game. It was an adult game based on Cream Lemon [a legendary pornographic animation series]. After my designs were used for *Always Together* (Zutto issho), people started to really like my characters and I've been working in the industry since.[10]

But times are tough for adult computer games, which are suffering from declining domestic sales. I asked Watanabe if he thinks that growing interest in the US might be a solution, to which he replied:

In truth, I don't want Americans to express interest. I don't want these games to be known. That is why I've refused all interviews on *otaku* culture [i.e., manga/anime subculture]. To start with, *otaku* culture isn't exactly a good thing. It's hidden in the back. In truth, I don't want it to come to the front. If the mass media keeps making a fuss over *otaku*, then many people will become aware of the culture. Adult computer games are not a good thing. Well, they are good, but not good, if you know what I mean. If these games come to the front, we can't feel affection for fictional characters. We can't do it. If *otaku* culture is too open, then our power of imagination and creativity will decline. […] Doing it in secret is Japanese culture.

Looking back at my field notes, I see that I was disheartened by the exchange and conflicted about what I, a cultural anthropologist, was doing in Japan with adult computer games. I saw Watanabe's point, but his radical rejection of outside interest was at odds with my commitment to encountering others and trying, impossible as it may be, to see and feel and move as they do. Idealism aside, I had to consider the politics. As someone working in the adult gaming industry, Watanabe's position against government and global interest makes sense: Adult computer games are not necessarily part of 'Cool Japan'; responding to global trends, the government is as likely to regulate the content as it is to promote it; and the population of players he imagines is a closed circle that already appreciates the content. My writing about adult computer games may draw unwanted attention that fuels Japan bashing and hurts the industry. Looking at *Popotan*, which features a character that appears very young and engages in explicit sex scenes, I appreciate that a higher profile might impact the creation and circulation of such games.

In the end, I decided to write this chapter, encouraged by earlier encounters with two women involved in the adult gaming industry and connected to Watanabe Akio. The first was Momoi Halko, who played these games in her youth, rose to fame singing the opening theme songs for them, and voiced a character in *Popotan*. Momoi, who profoundly respects Watanabe, differs from him, acting as something of an ambassador of Akihabara, manga/anime subculture, and affection for fictional characters, which she finds can connect people around the world (Galbraith, 2014: 72–9). When affection is open and shared, Momoi told me, it can change us.

The second source of inspiration was Toromi, who worked in the adult gaming industry, voiced a character in *Popotan*, and is now a character designer herself. At her studio in Hachiōji in the suburbs of Tokyo, I asked Toromi what she thought about the possibility that her manga/anime-style cute girl characters, so similar in their youthful sexuality to those of Watanabe, might spark outrage in global circulation. Toromi smiled and said, 'misunderstanding is part of being interested. It's nothing to be worried about. It's the beginning of a discussion, which may lead to understanding' (Galbraith, 2014: 88). Well said. But it helps if all parties involved acknowledge that they do not know as much as they might like to think. And that is not where things stand now, when so many seem so sure that they know about 'Japan'.

'As the gaming market grows', Consalvo writes, 'game makers will likely begin to tailor their products to highly individualized sub-cultures or groups, based not on nationality but instead on gender, age, and favourite hobbies' (2006: 133). This does seem accurate, but, in the case of the tailored products of adult computer games, there is nevertheless a push and pull of the content across national borders and a simultaneous renationalising of it as 'Japanese'. What Consalvo did not predict is that some tailored content, subcultures, and hobbies would become targets of criticism and regulation, or that this response would be renationalised in terms of 'us' and 'them'. Perhaps it is not that 'some games will never "translate" to other languages or cultures' (Consalvo, 2006: 133), but rather that there is sustained resistance to such translation; when they are translated, it is often in terms of difference, tinged with disgust about the deviant and dangerous others associated with 'pervert games from Japan'. The feedback in this transmission and resistance is the noise of discourse about 'Japan' (Novak, 2013: 10, 24). Following Iwabuchi, I would like to suggest the need to go beyond national and nationalising discourses to see the nuance of interactions within national borders and exceeding them

(Iwabuchi, 2010: 88, 93). Then, when someone such as Bamboo asks, 'Is this Japan?' we might hear more than a rhetorical question and think critically about shifting positional identities and relations in the world today. We might see emergent alliances and cleavages in and around 'Japan'. And we might grasp why someone such as Watanabe divides his world in terms of 'Americans' like me and 'Japanese' like him, even as this nationalising discourse obscures so much more than it reveals.

Notes

1   As Consalvo rightly states, 'we must ensure that examinations of the content of games does not slide too far into essentialist statements of national cultures' (Consalvo, 2006: 128). Here one might think of persistent discourse concerning the 'Japanese Lolita complex', which psychiatrist Saitō Tamaki flags as part of 'the miserable business of repetitive theorizing about the uniqueness of the Japanese' (Saitō, 2011: 7). 'Ethical considerations aside', Saitō explains, 'attributing a given perverse tendency to an entire nation based on a mere impression is simply an unscientific fallacy' (Saitō, 2011: 6–7). All the more so when one gets that impression from subcultural content such as adult computer games.

2   Theoretical discussions of circulation of media in general and pornography in particular provide useful orientation. In the field of porn studies, Susanna Paasonen has shown how the Internet raises questions about normative sexual and national boundaries (Paasonen, 2011), but the circulation of Japanese pornography in East Asia shines light on the complicated re-imaginings and re-creations of the nation. Be it mass circulation at the industry level in Taiwan (Wong and Yau, 2014), actress Aoi Sola using another language and social media to capitalise on 'rogue flows' in mainland China (Coates, 2014), or Chinese men desiring gayness as embodied by Koh Masaki (Baudinette, 2020), lines are drawn and crossed in pornographic texts and bodies, and Japan comes to matter somehow. Similar tension exists in discussions of pop imperialism and cosmopolitanism (Ching, 1996; Jenkins, 2006) and regional popular culture and structures of feeling (Lamarre, 2015; Otmazgin, 2016). As theorist Leo T.S. Ching rightly points out, important in all of this are imaginary relationships with 'Japan' (Ching, 2019; see also Galbraith, 2019). With this theoretical orientation in mind, building on concrete examples of encounters and practices, this chapter examines how and why 'Japan' comes to matter in the circulation of adult computer games.

3   This is the English translation of Nakasatomi's article, which was published in an online journal in 2013.

4   See the website for the event at http://www.excaddy.jp/charara/top.html.

5   See the J-List website at http//www.jlist.com.

6   See the Sekai Project website at https://sekaiproject.com/.

7   Unless otherwise indicated, all quotes from Bamboo in this section come from a personal interview on 6 October 2014.

8   See the MangaGamer website at http://blog.mangagamer.org/.

9   For a sense of the kind of space that this sharing opens, see this video introducing to the Hentai Festival in California: https://www.youtube.com/watch?v=NuuJy4bbFUc&mc_cid=4c5f05e6ff&mc_eid=4e834a4604+target%3D.

10 Unless otherwise indicated, all quotes from Watanabe in this section come from a personal interview on 23 May 2014.

References

Alexander, L. (2009) 'And you thought *Grand Theft Auto* was bad: Should the United States ban a Japanese "rape simulator" game?', *Slate*, 9 March, Available from: http://www.slate.com/ [Accessed 20 March 2020].

Allison, A. (2006) *Millennial Monsters: Japanese Toys and the Global Imagination*, Berkeley: University of California Press.

Baudinette, T. (2020) 'Aspirations for "Japanese gay masculinity": comparing Chinese and Japanese men's consumption of porn star Koh Masaki', *Porn Studies*, 7(3): 258–68.

Bogost, I. (2011) *How to Do Things with Videogames*, Minneapolis: University of Minnesota Press.

Ching, L. (1996) 'Imaginings in the empires of the sun: Japanese mass culture in Asia', in J.W. Treat (ed) *Contemporary Japan and Popular Culture*, Honolulu: University of Hawaii Press, pp 169–94.

Ching, L.T.S. (2019) *Anti-Japan: The Politics of Sentiment in Postcolonial East Asia*, Durham, NC: Duke University Press.

Coates, J. (2014) 'Rogue diva flows: Aoi Sola's reception in the Chinese media and mobile celebrity', *Journal of Japanese and Korean Cinema*, 6(1): 89–103.

Consalvo, M. (2006) 'Console video games and global corporations: Creating a hybrid culture', *New Media and Society*, 8(1): 117–37.

Consalvo, M. (2016) *Atari to Zelda: Japan's Videogames in Global Contexts*, Cambridge, MA: The MIT Press.

Coombe, R. (1998) 'Contingent articulations: A critical cultural studies of law', in A. Sarat and T.R. Kearns (eds) *Law in the Domains of Culture*, Ann Arbor: University of Michigan Press, pp 21–64.

Eng, L. (2012) 'Strategies of engagement: Discovering, defining, and describing otaku culture in the United States', in M. Ito, D. Okaben and I. Tsuji (eds) *Fandom Unbound: Otaku Culture in a Connected World*, New Haven, CT: Yale University Press, pp 85–104.

Equality Now (2009) 'Japan: Rape simulator games and the normalization of sexual violence', Equality Now, Available from: http://www.equalitynow.org/take_action/japan_action332 [Accessed 20 March 2020].

Fennelly, G. (2009) 'Exclusive: Amazon selling rape simulation game *Rapelay*', *Belfast Herald*, 12 February, Available from: http://www.belfasttelegraph.co.uk/ [Accessed 20 March 2020].

Galbraith, P.W. (2014) *The Moe Manifesto: An Insider's Look at the Worlds of Manga, Anime and Gaming*, Clarendon, VT: Tuttle Publishing.

Galbraith, P.W. (2017) '*RapeLay* and the return of the sex wars in Japan', *Porn Studies*, 4(1): 105–26.

Galbraith, P.W. (2019) *Otaku and the Struggle for Imagination in Japan*, Durham, NC: Duke University Press.

Game Faqs (2017) 'Why are Japanese games so focused on fetishism?', Available from: https://gamefaqs.gamespot.com/boards/691087-playstation-4/74805414?page=12 [Accessed 20 March 2020].

Game Politics (2009) 'In Parliament, suggestion of "global regulatory future" for video games', Available from: http://webmail.vgol.com/2009/07/21/parliament-suggestion-quotglobal-regulatory-futurequot-video-games [Accessed 20 March 2020].

Getchuya (2006) 'Nenkan sērusu rankingu' [Yearly sales ranking], Available from: http://www.getchu.com/pc/salesranking2006.html [Accessed 20 March 2020].

Grayson, N. (2019) 'Two of Steam's top games last month were anime sex games', *Kotaku*, 19 August, Available from: https://kotaku.com/two-of-steams-top-games-last-month-were-anime-sex-games-1837387431 [Accessed 20 March 2020].

Hall, S. (1998) 'Notes on deconstructing "the popular"', in J. Storey (ed) *Cultural Theory and Popular Culture: A Reader*, Upper Saddle River, NJ: Prentice Hall, pp 442–53.

Hebdige, D. (2005) 'Subculture', in R. Guins and O.Z. Cruz (eds) *Popular Culture: A Reader*, London: Sage, pp 355–71.

Hinton, P.R. (2014) 'The cultural context and the interpretation of Japanese "Lolita complex" style anime', *Intercultural Communication Studies*, 23(2): 54–68.

Iwabuchi, K. (2010) 'Undoing inter-national fandom in the age of brand nationalism', in F. Lunning (ed) *Mechademia 5: Fanthropologies*, Minneapolis: University of Minnesota Press, pp 87–96.

Jenkins, H. (2006) *Fans, Bloggers, and Gamers: Exploring Participatory Culture*, New York: New York University Press.

Jones, M.T. (2005) 'The impact of telepresence on cultural transmission through bishoujo games', *PsychNology*, 3(3): 292–311.

Kagami, H. (2010) *Hijitsuzai seishōnen ron: Otaku to shihonshugi* [On non-existent youth: Otaku and capitalism], Tokyo: Ai'ikusha.

Kelty, C. (2005) 'Geeks, social imaginaries, and recursive publics', *Cultural Anthropology*, 20(2): 185–214.

Lah, K. (2010) '"RapeLay": video game goes viral amid outrage', CNN [online], 31 March, Available from: https://edition.cnn.com/2010/WORLD/asiapcf/03/30/japan.video.game.rape.index.html [Accessed 4 January 2021].

Lamarre, T. (2015) 'Regional TV: Affective media geographies', *Asiascape: Digital Asia*, 2(1–2): 93–126.

Leonard, S. (2005) 'Progress against the law: anime and fandom, with the key to the globalization of culture', *International Journal of Cultural Studies*, 8(3): 281–305.

McLelland, M. (2005) 'The world of yaoi: The internet, censorship and the global "boys' love" fandom', *Australian Feminist Law Journal*, 23(1): 61–77.

Mainichi (2009) 'Seibōryoku gēmu: Rinri kikō ga seizō hanbai kinshi' [Sexually violent games: Ethics organization bans production and sale], Available from: https://web.archive.org/web/200906061425 28/http://mainichi.jp/select/wadai/news/20090604k0000m04007 8000c.html [Accessed 20 March 2020].

Mizuho Survey (2014) 'Kontentsu sangyō no tenbō: Kontentsu sangyō no sara naru hatten no tame ni' [Contents industry outlook: For the further development of the contents industry], *Mizuho sangyō chōsa*, 48(5): 1–149.

Nakasatomi, H. (2013) '"*RapeLay*" and the problem of legal reform in Japan', (C. Norma, trans), *Electronic Journal of Contemporary Japanese Studies*, 12(3), Available from: http://www.japanesestudies.org.uk/ ejcjs/vol12/iss3/nakasatomi.html [Accessed 20 March 2020].

Napier, S.J. (2007) *From Impressionism to Anime: Japan as Fantasy and Fan Cult in the Mind of the West*, New York: Palgrave.

Novak, D. (2013) *Japanoise: Music at the Edge of Circulation*, Durham, NC: Duke University Press.

Otmazgin, N. (2016) 'A new cultural geography of East Asia: Imagining a "region" through popular culture', *The Asia Pacific Journal: Japan Focus*, 14(7): 1–12.

Paasonen, S. (2011) *Carnal Resonance: Affect and Online Pornography*, Cambridge, MA: The MIT Press.

Pelletier-Gagnon, J. and Picard, M. (2015) 'Beyond *RapeLay*: Self-regulation in the Japanese erotic video game industry', in M. Wysocki and E.W. Lauteria (eds) *Rated M for Mature: Sex and Sexuality in Video Games*, London: Bloomsbury, pp 28–41.

Saitō, T. (2011) *Beautiful Fighting Girl* (J.K. Vincent and D. Lawson, trans), Minneapolis: University of Minnesota Press.

Sakakibara, H. (2016) 'Eroge gyōkai eroge no uriage ga nobinayamu igai na riyū towa' [Eroge business: The unexpected reason that adult computer games are struggling with sales], *Raku Job*, 23 January, Available from: http://raku-job.jp/news/companyrep/8705/ [Accessed 20 March 2020].

Taylor, E. (2007) 'Dating-simulation games: Leisure and gaming of Japanese youth culture', *Southeast Review of Asian Studies*, 29: 192–208.

Wong, H. and Yau, H. (2014) *Japanese Adult Videos in Taiwan*, London: Routledge.

Yano (2014) '"Otaku" shijō ni kan suru chōsa kekka 2014' [Survey results on the 'otaku' market: 2014], Yano Keizai Kenkyūjo, Available from: https://www.yano.co.jp/press/pdf/1334.pdf [Accessed 20 March 2020].

PART II

Governance and Regulations

Introduction to Part II

Micky Lee

The second part contains four chapters about digital media governance. We begin with Ana Gascón Marcén's chapter on data privacy protection in Japan and South Korea, followed by Deirdre Sneep's chapter on government warnings about playing *Pokémon GO* in public places. Both chapters look at regulations of digital technologies at the national level. The next two chapters focus on North Korea: Micky Lee and Weiqi Zhang discuss why the Western concept of intellectual property does not apply to counterfeit media content in the country, while Elizabeth Shim discusses how the North Korean government borrows images from South Korean popular culture imagery to showcase its nuclear weapons to citizens and the world. While the first two chapters examine digital technologies in East Asia from a regulatory perspective, the next two use governance as a jumping board from which political-economic and cultural critiques of digital technologies are effected.

   This introduction does not aim to provide an exhaustive account of how the three countries respond to technological advancement and international pressure. It will however point out how the four chapters illustrate two tensions in regulating technologies at the national level: First, national governments take into account international governance of digital technologies; second, the 'objects' that need to be regulated are often deemed harmful or undesirable to citizens even though they find loopholes in the regulations and use banned technologies behind the back of the state. These two tensions illustrate the dynamic relationship between structure and agency: To what extent do East Asian governments have complete agency to shape international regulations or do international standards constrain national policies? Can national regulations change how citizens conceptualise technologies or does active use of technologies shape regulations? In addition, humans are not the only actors who have agency (Latour, 2005), technologies also have agency to shape how social actors interact with them. As a result, even though governments try to control how technologies are used, technologies also enable users to do things that governments cannot control. The dynamic relationship between structure and agency once again affirms why this edited book uses three approaches – political economy, critical cultural studies, and science and technology studies – to examine media technologies in East Asia.

Broadly speaking, regulations of digital technologies can be categorised into those of ownership and those of content or data. Regulations of ownership include who can invest in technologies as well as who can own intellectual property. A country's political-economic system determines who – public or private firms – can invest in technology development and own intellectual property. For example, in Japan and South Korea, corporations are private firms that produce and distribute technologies, therefore they own the intellectual property of the invention. In contrast, North Korea has more state-owned companies even though it is unclear how much digital technologies are invented in the country. Regulations of content/data include who can produce, share, and consume what kinds of content and data. A country's political-economic system once again determines what kinds of content and data are permitted for sharing. For example, the democratic political system of Japan and South Korea allows citizens and journalists to criticise political leaders; doing so in North Korea may attract the death penalty. Similarly, democratic states have more concern for personal liberty and may not collect all information about its citizens. In contrast, the authoritarian state of North Korea wishes to know every move and thought of its citizens so it collects all conversation that citizens have on cellphones.

Even though ownership regulations can be separated from content regulations, the two often influence each other. For example, corporations want to regulate the circulation of media content because they own the intellectual property of the content. In the case of North Korea, because media are not seen as money-making, regulators do not allow for certain kinds of content even if there is a market demand for it.

Ana Gascón Marcén examines how the Japanese government developed national policies of personal data protection. Its government did so not only to respond to scandals in the country, but also to submit to international pressure, in particular from the EU. However, Japan could not directly borrow international regulations of personal data protection because the local culture does not have the concept of 'privacy'. Therefore, the government has to balance both international and national interests when designing data protection laws. The international community also exerts pressure on countries that do not enforce intellectual property laws. For example, the US has been complaining about counterfeit Hollywood DVDs at international forums such as the World Trade Organization and the World Intellectual Property Organization. However, when counterfeit media is widely bought and sold in the black market in North Korea, the Western media are not too concerned with intellectual property violations, instead they believe that such efforts will bring positive change to North Korean society.

Micky Lee and Weiqi Zhang look at how foreign media were brought into the country and how the Western media turned a blind eye to the pervasive intellectual property violation in North Korea. The Western media believe that foreign media may unleash North Korean citizens' desire for freedom and democracy, which may eventually bring capitalism to the country and integrate it into a global economy. Therefore, losing profits from counterfeit media is a small price to pay if the North Korean state eventually adheres to more international standards, including humans rights and a free-market economy.

The second tension that the four chapters illustrate is whether regulations can control technologies or not. The objects that need to be regulated are often deemed harmful to the society, therefore laws and ordinances would mitigate the harm. However, regulations are often ineffective at controlling users' behaviours. In fact, unexpected ways to use technologies sometimes have desirable effects in society.

Deirdre Sneep examines the negative view that the media have towards the mobile game *Pokémon GO*. Health and public safety results were cited to explain why this game could bring harm to Japanese society. Accordingly, the government also issued warnings about chronic playing leading to health problems and car accidents. Sneep has shown that *Pokémon GO* is not the first technological device that is believed to bring social harm, the Sony Walkman had also been subject to such criticism. The regulations seem to reveal a fear that technologies – as new social objects – will harm traditional social fabrics; technologies are believed to eradicate a respect for the collective when members indulge in too many individualistic activities. Yet, some gamers have gained unexpected benefits such as getting out of the home to exercise and meeting new people.

In North Korea, foreign media have been deemed harmful to society; foreign media are seen as pollutants that threaten the superiority of the Korean race. The government strictly regulates what media content the populations will see. Therefore, the North Korean state owns all media, it eavesdrops on messages on mobile phones and limits Internet access to only a few hundred people. Yet, these regulations have not stopped the populations from selling, buying, and consuming foreign media. As previously mentioned, some believe that the circulation of foreign media – despite pervasive intellectual property violations – may bring benefits to the populations because the North Korean public may have a taste of freedom and democracy in a restrictive society.

Another unexpected benefit of pirated media is the appropriation of spectacular images for mass persuasion. As Elizabeth Shim shows, North Korean leaders counter the alluring images of South Korean popular culture by staging their televised nuclear weapon shows. Because images of North Korean nuclear capacity all come from state-owned television, the actual quantity and power of North Korean nuclear weaponry are unknown. Shim argued that these images create a hyperreality of North Korea, that any representation of North Korea is less a reflection of the reality but more an image that the North Korea state wants North Korean citizens and the world to see. In a way, these images share similarities with representations of South Korean popular culture because they are circulated in image-saturated media.

Reference

Latour, B. (2005) *Reassembling the Social: An Introduction to Actor-network Theory*, Oxford: Oxford University Press.

3

The New Personal Data Protection in Japan: Is It Enough?

Ana Gascón Marcén

Introduction: The importance of personal data protection in the digital age

The protection of personal data is extremely important for work and play in the digital environment. The ubiquitous use of information and communication technologies results in a constant flow of data and the transformation of social action into online quantified data that allows for real-time tracking and predictive analysis (Mayer-Schoenberger and Cukier, 2013). Many digital services seem free of charge to the users, but they are financed through the commodification of personal data, which has become a common business model for digital companies. Data is a regular currency for users to pay for 'free' services. As van Dijck (2014) remarked, this has become a trade-off nestled into the comfort zone of most people. This helps to fuel what Zuboff (2019) has defined as 'surveillance capitalism', a new economic industry deeply rooted in our society, which exploits users' data to predict and control human behaviour.

With the development of greater computing power, huge amounts of data are collected and processed to obtain useful information and accurate profiles of individuals. In addition, the Internet of Things –

the installation of sensors in a wide variety of objects interconnected through the Internet where the objects receive and send information autonomously – and the use of artificial intelligence will significantly increase the trend of personal data collection.

There are huge advantages to these technologies. For example, they can lead to better medical treatments, more efficient public transport systems, and better-managed stocks for companies. However, there is also a dark side to these advancements: there is a large amount of personal data at the disposal of governments and private businesses. Having access to this information consolidates power in the hands of these actors. This information has an enormous impact on the workplace, because all these data could be used to streamline procedures and assess productivity, but they can also be essential to the hiring or even the firing process, when they become automated (Jee, 2019). Cases have been reported in Japan in which companies analyse the data of their workers (including from their social networks) through algorithms to calculate the probability that they will steal from the business. Data are also used to analyse if employees are likely to share confidential information outside the company or to behave inappropriately at work (Fitzpatrick, 2015).

After the Cambridge Analytica–Facebook data breach scandal, it became clear that companies' personal data collection can affect even the very fabric of democratic societies, as companies with enough personal data and the ability to generate insights from them can influence how millions of people vote.

Every individual has a digital footprint because online interactions leave a trail and the data are aggregated. On the one hand, there are users who put a great amount of effort into shaping their digital identities. On the other hand, some people are not even aware of their digital footprint, because of lack of education or resources. This gap of information literacy may further entrench social inequalities (Micheli et al, 2018). Past activities that are irrelevant to the present can persecute users forever if their information appears in online search results. The digital footprint has given rise to concepts such as digital self-determination or the right to be forgotten.[1]

The implications of digital footprints are especially important in Japan, which has one of the highest percentage of Internet users: Around 91 per cent of the population was online in 2017 (International Telecommunication Union, 2018). Japan has a strong digital infrastructure and the government invested heavily in wifi networks ahead of the Tokyo Olympics, that had been planned for 2020.

Because of the pervasive use of the Internet, there is a pressing need to regulate the scope of personal data protection; the exchange of data involves many risks, but in turn can lead to important technological advances. Therefore, it is imperative to establish what principles should govern data processing.[2]

Therefore, this chapter will analyse how the concept of personal data protection has evolved in Japan and the current legal regime. It will be assessed from a European perspective as that is the region with the highest level of protection worldwide. The chapter will also compare the Japanese approach with that of its neighbouring country South Korea, as both countries are democracies that decided to enhance their personal data protection regime in order to be recognised by the European Commission as having an equivalent level of protection as the European Union (EU), but they chose different paths to do it. It is not a three-way comparison as much as an analysis of how Japan changed its regulations to satisfy the EU's requirements. The case of South Korea provides some additional information about how another East Asian country responded to the EU.

The protection of personal data in Japan: The creation of a new concept

Traditionally, there has been a lack of awareness of the importance of privacy in Japan. This is because historically Japanese society was rather collective (Nakada and Tamura, 2005; Orito and Murata, 2005), something reinforced by the *tonarigumi* or neighbourhood associations before the US occupation. In fact, there is no precise Japanese translation for the word 'privacy'. Many Japanese used the word プライバシー (*puraibashī*), a word imported from English, without understanding its full meaning (Murata, 2004). It is a concept that has evolved over the years, taking in Western elements, but also Japanese ones, such as respect for the relationship between oneself and others (Miyashita, 2011a). The Japanese Constitution does not specifically recognise the right to privacy; however, the Supreme Court has protected it since the 1970s, based on Article 13 of the Constitution that establishes the right to life, freedom and the pursuit of happiness.

The issue of personal data protection only gained steam in the late 1990s, although there were already some precedents.[3] At that time, the external pressure increased since the European Community made the flow of personal data to countries outside the European Economic Area conditional on an adequate level of data protection in the destination country. That is why Japanese companies were convinced of the country's need to draft data protection regulations (Orito and Murata, 2005).

Another factor that led to enhanced data privacy was the modification of the census of residents, creating a network in 2002 called *Juki-Net* that recorded names, addresses, dates of birth, and genders. The purpose of this network was to facilitate the exchange of data between municipalities when residents moved from one to another, but it raised serious doubts about its impact on citizens' privacy. The Supreme Court decided in 2008 that the collection of citizens' data was legal and that all citizens should be included (Masao, 2012).

The Diet passed the first Japanese law on personal data protection in May 2003, the Act on the Protection of Personal Information (APPI), which began to create obligations for companies in 2005. The main novelty of the APPI was that it applied to the private sector (previously self-regulated), specifically to all companies that had processed data of more than 5,000 individuals in the previous six months. Each ministry that regulated specific industries then approved guidelines for its application in each sector.

In order to improve compliance, the APPI established sanctions of up to six months in prison and fines of up to 300,000 Japanese yen. However, the most powerful reason that drove Japanese companies to comply with the APPI was that they did not wish to have their reputations harmed by violating the trust of their customers, as this is viewed extremely negatively in Japanese culture (Orito and Murata, 2008). In fact, in Japan, when there are personal data breaches, it is usual for companies to voluntarily pay compensation to users and ask for forgiveness.

The APPI relied heavily on the concept of the purpose of data processing. Each company had to advertise exactly how it processed the data according to the pre-established purpose. The emphasis on this criterion is different from the regimes of the European countries, for example (Shoniregun et al, 2010). Under the APPI, data subjects can request that a business disclose how their personal data will be used, how they can access or correct their data, and where they can submit complaints. Data subjects can also demand that an organisation correct their personal information or delete it if it is not used according to the stated purpose, is transferred without prior consent, or was acquired by fraud or other unfair means. According to the APPI, companies were obliged to prevent the unauthorised disclosure, loss, or destruction of personal data and only transfer data to third parties with the consent of the individuals concerned.

One of the criticisms of the APPI was that companies need to inform individuals about the 'opt-out' schemes, but these mechanisms were never practically implemented. Another criticism was that sometimes the stated purpose for which the data was used was overly vague (Orito and Murata, 2008).

The reform of the Act on the Protection of Personal Information

Almost a decade after the APPI came into force, it became outdated. The same reasons that led to its drafting made its reform imperative: an international trend of modernisation, and information and communication technologies advancement. The redrafting of the APPI was an opportunity to promote technological innovation. Although Japan had one of the best communications infrastructures, it was lagging in the research and commercial application of new technologies that make intensive use of data. Finally, another trigger for the modernisation of the law was the scandals linked to personal data leaks and the development of a new personal record with all the risks that this could entail.

In 2013, the Diet passed the Act on the 'Use of Numbers to Identify a Specific Individual in the Administrative Procedure' (known as "My Number"). Identification numbers were introduced in 2015 for all residents of Japan, including foreign nationals. This number is linked to very sensitive data related to social security and taxes. Its use became mandatory from 1 January 2016. 'My number' allows access to certain public services and related procedures, such as pensions and unemployment benefits. Furthermore, since January 2016, 'My Number' is listed on the withholding slip and various other payment records of salaried employees.

Politicians and the administration strongly emphasised the advantages 'My Number' would bring, such as increasing administrative efficiency, reducing bureaucratic procedures, fighting fraud, identifying natural disasters victims, and so on. However, this initiative also poses risks because of the possibilities of personal data leaks and identity theft as a result of cyberattacks. Those who have criticised the possible problems of the norm are not cybersecurity paranoids but professional associations with a thorough knowledge of the subject matter, such as the Japanese Federation of Bar Associations and the Japanese Medical Association (Tkach-Kawasaki, 2015). Currently, only administrative agencies and employers can use 'My Number' because of the fear of possible misuse by the private sector. Nevertheless, in the future, the plan is to use 'My Number' to access different services like logging in to Internet bank accounts or ticketless services (My Number Promotion Office, 2015).

In recent years, in Japan, several scandals have occurred that involved the leak of sensitive data of millions of people. In 2014 Japan Airlines reported that its points programme for frequent travellers had suffered a cyberattack that could have affected the data of up to 750,000 of its customers (Horowitz et al, 2015). However, this security breach was quickly overshadowed by the one involving Benesse, an education company. One of its engineers sold the data of more than 30 million people. It is not only companies that have had security breaches, government agencies have also faced the same problem. In 2015, there was a leak of the personal data of 1.25 million people due to a failure of the Japanese pension system (Mallard and Sieg, 2015). In 2013, the government decided to initiate a sweeping reform of the APPI, which was promulgated in 2015. The amendment expanded the definition of personal data, extending it to biometric information such as fingerprints or data related to facial recognition.[4] Furthermore, the category 'special care-required personal information' refers to personal information comprising a subject's race, creed, social status, medical history, criminal record, record of being a crime victim, and other personal information with the potential to create unjustifiable discrimination or prejudice.

Another substantial modification relates to the anonymisation of data that may be transferred without the consent of its subject. This was announced as one of the main advantages of the amendment, as it allowed Japanese companies to innovate and make greater use of big data (Miyashita, 2015). However, it is extremely complex to design effective anonymisation systems that are impossible to reverse, because the crossing of different databases creates risks of re-identification (Article 29 Working Party, 2014).

One of the most positive changes of the amendment of the APPI is the creation of a truly independent and centralised data protection authority. Japan was one of the few countries with a personal data protection law that did not have an independent authority. This does not mean that there were no public bodies responsible for such protection; different ministries were dedicated to specific areas, and the supervision of the application of the APPI was entrusted to the Consumer Protection Agency as of 2009 (Miyashita, 2011b). However, this solution was still unsatisfactory since the agency was not an independent body dedicated exclusively to data protection and lacked many of the competencies that data protection authorities usually have.

When 'My Number' was created, it was decided that an independent body should supervise its use. Thus, in 2014 the Specific Personal Information Protection Commission was launched. This commission was responsible for ensuring the correct use of 'My Number' and other specific personal information and issuing recommendations (Miyashita, 2014). Unfortunately, it lacked the necessary means to ensure compliance with the regulations, although it could request reports and carry out onsite inspections of companies.

Despite its limited scope of action, this commission became the seed of an authentic data protection authority, the Personal Information Protection Commission (PPC) created by the 2015 amendment to the APPI and established on 1 January 2016. This independent commission consists of a chairperson and eight members. It is the central authority in charge of ensuring that the private sector complies with regulations on personal data protection, with strengthened powers and resources. Its duties are the protection of the rights and interests of individuals while taking into consideration proper and effective use of personal information. It can receive complaints, perform audits, and issue cessation orders (Greenleaf, 2015). One of its main competencies is the elaboration of the regulations for the development of the APPI, including those related to specifying anonymisation processes, and the preparation of the list of states with a sufficient level of protection to be able to make personal data transfers.

The amendment meant that the APPI covered all private companies eliminating the previous exemption for companies that processed data of fewer than 5,000 individuals. In addition, criminal penalties were created that can reach up to one year in prison or fines of up to 500,000 Japanese yen for those who disclosed personal data for illegal purposes.

The amended APPI included some restrictions regarding the transfer of personal data to other countries, for example, requiring prior consent of the data subject to do so except where the third party is in a state where regulation on personal information protection is equivalent to that of Japan.

Japanese and South Korean data protection laws from a European perspective

Both Japan and South Korea are important trade partners of the EU, which has the highest level of personal data protection in the world because of its General Data Protection Regulation (GDPR).[5]

The EU–South Korea Free Trade Agreement was provisionally applied since 2011 and formally ratified in 2015, while the EU–Japan Economic Partnership Agreement entered into force in 2019. To fulfil the potential of both agreements, it would be advantageous to make the transfer of personal data with the EU easier. For this purpose, the European Commission has the power to determine whether a country outside the European Economic Area offers an adequate level of personal data protection, essentially equivalent to that offered by the GDPR, through and adequacy decision. This follows the same logic that was included in the 2015 reform of the APPI, as explained already.

The European Commission announced in 2017 that it would actively engage with key non-EU countries to explore the possibility of adopting adequacy findings, starting with Japan and South Korea, with a view to fostering regulatory convergence with European standards and facilitating trade relations (European Commission, 2017a).

The issue was that both South Korean data protection laws and the Japanese amended APPI, even though converging with the GDPR, did not offer the same level of protection. After the assessment, South Korea and Japan opted for different approaches. While South Korea decided to amend its laws dealing with personal data protection, Japan preferred not to modify them but instead took measures of another kind. In the following section, the Japanese solution will be analysed first, preliminary to comparing it with the Korean option.

The Japanese path towards an adequacy decision

The 2015 amendment to the APPI was motivated partly by the objective of achieving an adequacy decision from the European Commission. However, the modification did not go far enough to attain it. Nevertheless, from both the European and Japanese perspectives, there was interest to get such a decision, so a creative solution emerged. This was not the first time that the European Commission had to opt for a partial and convoluted way to reach such a decision. It was also the case with the US, because the flow of personal data between the EU and the US is paramount for both economies, but the US did not offer equivalent protection to personal data. That was the reason for the creation of 'Safe Harbour', a mechanism that eased the transfer of personal data between the European Community and companies in the US that self-certified their adherence to certain principles of personal data protection. Nevertheless, the Court of Justice of the EU invalidated it (the *Schrems* decision) as well as its successor the 'Privacy Shield' (the *Schrems II* decision).

In the case of Japan, a different solution emerged because Japanese law created a legal environment that was much closer to what the EU was looking for than the US did. Therefore, only a few additional safeguards were needed. The Japanese government delegated to the

PPC, as the authority administering and implementing the APPI, the power to take the necessary action to bridge the differences of the systems to ensure appropriate handling of personal information received from Europe. Therefore, the PPC in June 2018 adopted the 'Supplementary Rules under the Act on the Protection of Personal Information for the Handling of Personal Data Transferred from the EU based on an Adequacy Decision' with a view to enhancing the protection of personal information transferred from the EU to Japan. Those Supplementary Rules are legally binding on Japanese business operators and enforceable, both by the PPC and by courts, in the same way as the provisions of the APPI that they supplement with stricter and more detailed rules.

A very clear example to show how this works is sensitive data. The PPC decided that the protections afforded to 'special care-required personal information' under the APPI were extended to all categories recognised as 'sensitive data' in the GDPR through the Supplementary Rules. The APPI does not protect data regarding sexual activities, sexual orientation, or trade union membership, for example, but in European standards, this is considered very sensitive information.

The idea of equivalent protection not only covers what the laws stipulate, but also their effective enforcement, so the Japanese government also promised the creation of a new redress mechanism, administrated by the PPC, to handle complaints by individuals residing in the EU concerning access of Japanese entities to their personal data transferred from the EU.

It is necessary to underline that the adequacy decision only applies to the personal data transferred from the EU to businesses in Japan, but not to the public sector; still there were doubts of what would happen if Japanese public authorities requested access to these data from businesses for the purposes of criminal law enforcement or national security. In this regard, the Japanese government gave a series of assurances and made commitments to the European Commission to ensure that any interference with the fundamental rights of the individuals whose personal data are transferred from the EU to Japan by Japanese public authorities for public interest purposes will be limited to what is strictly necessary to achieve the legitimate objectives in question, and that effective legal protection against wrongful interference exists.

The European Commission presented its draft adequacy decision in 2018. The European Parliament approved a Resolution on the adequacy of the protection of personal data afforded by Japan. It took note of the amendments to the APPI that entered into force in

2017, welcomed the substantive improvements and considered that the Japanese and EU data protection systems shared a high degree of convergence in terms of principles, safeguards, and individual rights, as well as of oversight and enforcement mechanisms. Nevertheless, it noted with concern some gaps and called on the European Commission to provide further evidence and explanation regarding several matters, in order to demonstrate that the Japanese data protection legal framework ensures an adequate level of protection that is essentially equivalent to that of the European framework.

The European Parliament raised the issue of media reports about the Japanese Directorate for Signals Intelligence, which allegedly employs about 1,700 people and has 'at least six surveillance facilities that eavesdrop around the clock on phone calls, emails, and other communications' (Gallagher, 2018). The European Parliament was worried that this element of Japanese indiscriminate mass surveillance was not even mentioned in the draft and that this mass surveillance would not stand the test of the criteria established by the Court of Justice of the EU in the *Schrems* judgment.

Another question is whether companies in Japan have enough incentives to comply with the law, as the fines for violating the APPI were extremely low compared with the ones for GDPR violations. The GDPR establishes that, for the most serious infringements, the fines are up to €200 million or, in the case of an undertaking, up to 4 per cent of the total worldwide annual turnover of the preceding financial year, whichever is higher.

The European Data Protection Board (EDPB) approved its Opinion 28/2018 regarding the European Commission Draft Implementing Decision on the adequate protection of personal data in Japan. The EDPB noted that there were key areas of alignment between the GDPR framework and the Japanese one on certain core provisions, such as data accuracy and minimisation, storage limitation, data security, purpose limitation, and an independent supervisory authority. Nevertheless, the EDPB said that further clarification was still needed on some aspects of the draft adequacy decision. These related, for example, to some key concepts of the Japanese legislation. More specifically, there was a lack of clarity around the status of the so-called 'trustee', a term that resembles the one used for the data processor under the GDPR but whose ability to determine and change the purposes and means of processing of personal data remained ambiguous. The EDPB also needed assurances on whether the restrictions to the rights of individuals (in particular, rights of access, rectification, and objection) were necessary and suitable for the legitimate aim pursued in a democratic society to make sure that they respect human rights.

In conclusion, the EDPB welcomed the efforts made by the European Commission and the PPC to align as much as possible the Japanese legal framework with the European one; however, the EDPB noticed that a number of concerns, coupled with the need for more clarifications, remained. Furthermore, this specific type of adequacy, combining an existing national framework with additional specific rules, also raised questions about its operational implementation. The EDPB recommended that the European Commission addressed the concerns and requests for clarification, and provide further evidence and explanations regarding the issues raised.

The Commission took into account some particular points of the opinion, for example, conducting a review of this adequacy finding every two years (not every four years as suggested in the draft). Nevertheless, it did not restart negotiations with Japan, so most of the demands for clarification were not met. The Commission probably was under a lot of pressure to approve the decision as quickly as possible, to coincide with the entry into force of the EU–Japan Economic Partnership Agreement.

Finally, the Commission approved Implementing Decision (EU) 2019/419 of 23 January 2019 pursuant to Regulation (EU) 2016/679 of the European Parliament and of the Council on the adequate protection of personal data by Japan under the Act on the Protection of Personal Information, which entered into force on the day of its adoption. The PPC also recognised the same day that the EU had equivalent standards to those in Japan regarding the protection of personal data. These could be considered mutual adequacy decisions. The EU and Japan boasted subsequently of having created the world's largest area of safe data flows. However, all this pressure to hasten decisions may have had a negative impact on the negotiations (Kanetake and de Vries, 2018).

In any case, it was not good for the Japanese data subjects as the new safeguards created by the Supplementary Rules only apply to the personal data coming from the EU. From a purely personal data protection perspective, it would seem that a more comprehensive approach that also grants those rights to people in Japan would have been better (Greenleaf, 2018a).

The APPI was partially reformed again in 2020: the amendment reinforced the rights of the data subjects; for example, it would have been easier for them to ask companies to delete or stop using their data, when there is a possible violation of their individual rights or they can show a legitimate interest. In addition, data subjects get to choose the methods in which the companies have to disclose to them the personal data that companies keep related to them. Furthermore, the law will apply to personal data stored for less than six months, because, until the amendment, there was an exception and it was not covered. The amendment also mandates companies to report to the PPC and notify data subjects when there is a leakage of their personal data that may violate individual rights and interests. An added safeguard is that, when personal data is shared with a third party outside Japan, the provider of personal data in Japan has to supply the data subject some basic information on the personal data protection system of the foreign country (PPC, 2020). The amendment also increases the possible fines in case of infringement of the APPI, the maximum for legal entities is 100 million yen, which is still quite low in comparison with the thresholds of the GDPR. The amendment is timely, as the European Commission has to review its adequacy decision in 2021.

The South Korean path towards an adequacy decision

An interesting, although lengthy, approach to getting an adequacy decision is the one adopted by South Korea. The South Korean government began negotiations with the European Commission as early as 2015. It gained traction in 2017 when both the Korean and European counterparts aimed for an adequacy decision in 2018 (European Commission, 2017b), but it did not happen as South Korea was still reforming its data protection legislation.

South Korea developed separate laws to regulate the use of personal data in the public and private sectors; the Act on Protection of Personal Information of 1995 applied to public institutions while the Act on Promotion of Information and Communications Network Utilisation and Information Protection of 1999 applied to the private sector. The current comprehensive Personal Information Protection Act (PIPA) enacted in 2011 integrated the two sectors. The Korean personal data protection regime is considered one of the strictest in the world. Like the GDPR, it protects privacy rights from the perspective of the data subject (Wall, 2018).

Greenleaf and Park (2014) argued that the principles in Korean law are clearly the strongest in Asia (although they were not yet fully enforced) and they even declared that, in some respects, the Korean principles go beyond those found in European laws, and indicate that innovation in data privacy legislation no longer originates solely in Europe. Nevertheless, it also has its critics, such as Ko et al (2016), who suggested that the Korean approach has not yet provided clear and predictable legal and practical standards for commercial actors in fields that rely upon data collection, processing, and sharing. Weighing its advantages and disadvantages, they consider that, while the laws may limit the emergence of data-based industries, it may not provide better protection of personal information for individuals.

Even with such a strict regime, some differences with the GDPR stood between Korea and the adequacy decision, especially the lack of independence of its main data protection authority and the fragmentation of the enforcement powers. The PIPA created a Personal Information Protection Commission (PIPC) that was a data protection authority with some independence but without enforcement powers of its own (Meyer, 2018).

In February 2020, South Korea enacted amendments to the three major data protection laws: the PIPA, the Act on the Promotion of Information and Communications Network Utilisation and Information Protection, and the Act on the Use and Protection of Credit Information. These amendments came into force in August 2020. The amendments aimed primarily at minimising the overlapping data privacy regulations and multiple supervisory bodies. They developed a 'data economy' by introducing the concept of 'pseudonymised data' and a legal basis on which data may be used more flexibly to an extent reasonably related to the original purpose of collection (Kang and Kim, 2020).

South Korea made the PIPC a 'central administrative agency' that is truly independent and with the authority over all situations of processing personal information, including enforcement and supervision. In this way, it will be empowered to investigate violations and to impose fines of up to 3 per cent of turnover (Greenleaf, 2018b).

These changes and others that were already approved brought South Korea even closer to the GDPR, for example in how it safeguards the cross-border transfer of personal information by restricting the onward transfer of personal information and making necessary the designation of a local representative (Park et al, 2018).

The only problem is that another motivation for the amendment to the PIPA was to deal with big data, as was the case in Japan. To make the processing of big data easier, the bill provides that a controller may process pseudonymised information without the consent of the data subject for the purpose of statistics, scientific research, public interest archiving, and so on.[6] This may not sit well with the GDPR, although some Japanese dispositions with the same aim were not a barrier for an adequacy decision. Nevertheless, some authors criticise the Korean regulations for their strong focus on data protection but insufficient consideration for facilitating data usage (Kim et al, 2018).

The forecasts of an adequacy decision in 2018 were wrong. However, it is quite probable that South Korea could get one in 2021, thanks to the amendment of the PIPA and the creation of a fully independent PIPC with enforcement powers.

Conclusion

In a society with pervasive digital technology in every aspect of our daily lives and constant Internet connections, it seems increasingly necessary to strengthen privacy to ensure effective protection of human rights. Citizens need to make responsible choices for themselves and be aware of the consequences of their actions. It is therefore interesting to see the evolution of this concept in Japan, from its appearance because of international pressure to its assimilation, adapting it to its own cultural characteristics.

The amendment of the PIPA in 2015 was imperative since it had been one of the weakest laws in Asia until its reform. One of the most positive developments was the creation of an independent personal data protection authority that oversees the compliance of the private sector. Its creation aligned Japan with the rest of the international community. The 2020 amendment has further improved data protection in Japan.

Nevertheless, national reforms are not enough for the improvement of personal data protection worldwide. The ideal situation would be an international treaty to regulate personal data protection adopted by the United Nations. However, this has proven to be extremely difficult. That is why South Korea and Japan are working in other forums, such as the Organisation for Economic Co-operation and Development (OECD) and the Asia-Pacific Economic Cooperation (APEC). They are also following closely the advances of the Council of Europe, becoming Observer States to the Committee of the Convention for the Protection of Individuals with regard to Automatic Processing of Personal Data (Gascón Marcén, 2016). Further co-operation with the EU will also be necessary to bring these standards to international trade debates such as the ones taking place in the World Trade Organisation (Gascón Marcén, 2020).

However, if Japan wants to be a leader in this field it has to do more. A step in the right direction would have been to apply the additional safeguards created through the Supplementary Rules to all personal data in Japan, not just those coming from the EU. Unfortunately, Japan opted for the narrowest possible path to adequacy depriving the

Japanese of its advantages (Greenleaf, 2018a). The 2020 amendment partially remedied this, but there are still safeguards that have to be put in place. In this sense, the South Korean approach to overhaul legislation seems slower but better for the interests of its citizens.

South Korea and Japan also share the desire to protect personal data while at the same time making large quantities of data available to the private sector to develop big data and artificial intelligence applications, as they do not want to lag behind China or the US in innovation. This will need further careful development of their legal frameworks in the coming years.

Notes

[1] The 'right to be forgotten' is derived from a decision of the Court of Justice of the EU in the case *Google Spain* (Judgment of 13 May 2014, *Google Spain SL and Google Inc.* v. *Agencia Española de Protección de Datos (AEPD) and Mario Costeja González*, C-131/12). The court ruled that the operator of a search engine is obliged to remove from the list of results displayed in response to a search of a person's name links to web pages when the data subject makes such a request, even if some exceptions have to be considered, such as when the information has public interest. This right has been widely criticised because it places such decisions in the hands of Internet intermediaries instead of judges and because it creates conflicts with the right to freedom of expression, including the right to information. Nevertheless, the General Data Protection Regulation of the EU enhanced it, creating a 'right to erasure' in Article 17.

In Japan, it has been quite controversial, as such kind of requests are usually declined by search engines and the affected persons have taken their cases to court. In 2017, the Supreme Court of Japan did not support such de-referencing requests in a case related to an arrest on sex charges (Judgment of the Supreme Court, 1 December 2017, case number 2016 (Kyo) 45, Minshu Vol. 71, No. 1). Nevertheless, other requests may succeed in different circumstances as the Supreme Court stated that the deletion of references could be allowed when the value of privacy protection clearly outweighs that of information disclosure. The court mentioned some criteria that are relevant to the decision on such requests and to finding the right balance, such as the nature and details of the facts, the degree to which the person suffers concrete damage, the person's social status and influence, the purposes and meaning of the website articles, the social situation at the time the articles were published, the social changes afterwards, and the need for including the relevant facts in the articles (Kanetake and Taylor, 2017).

[2] In the context of this chapter, 'processing' is used in a broad sense (as in the GDPR). It means any operation or set of operations performed on personal data or on sets of personal data, whether or not by automated means, such as collection, recording, organisation, structuring, storage, adaptation or alteration, retrieval, consultation, use, disclosure by transmission, dissemination or otherwise making available, alignment or combination, restriction, erasure, or destruction.

[3] In 1988, an act was passed that applied to the data stored in the computers of public administration agencies and, in some cases, local regulations were also developed. Regarding companies, for a long time, there were only some non-

binding guidelines prepared by the Ministry of Commerce and Industry for the private sector in 1989 (Fischer-Hübner, 1997).

4   'Personal information' in the APPI means information relating to a living individual which falls under any of the following items: those containing a name, date of birth, or other descriptions that are stated, recorded, or otherwise expressed using voice, movement, or other methods in a document, drawing, or electromagnetic record; or whereby a specific individual can be identified (including those which can be readily collated with other information and thereby identify a specific individual); or those containing an individual identification code.

5   Regulation (EU) 2016/679 of the European Parliament and of the Council of 27 April 2016 on the protection of natural persons with regard to the processing of personal data and on the free movement of such data and repealing Directive 95/46/EC (General Data Protection Regulation).

6   'Pseudonymisation' means the processing of personal data in such a manner that the personal data can no longer be attributed to a specific data subject without the use of additional information. Such additional information needs to be kept separately and subject to technical and organisational measures to ensure that the personal data are not attributed to an identified or identifiable natural person. It is different from 'anonymisation' because using additional information it can be reversed, which should not be the case with an effective anonymisation procedure.

References

Article 29 Working Party (2014) 'Opinion 05/2014 on Anonymisation Techniques', Available from: https://ec.europa.eu/justice/article-29/documentation/opinion-recommendation/files/2014/wp216_en.pdf [Accessed 5 January 2020].

European Commission (2017a) 'Communication to the European Parliament and the Council "Exchanging and Protecting Personal Data in a Globalised World"', COM/2017/07 final, Available from: https://eur-lex.europa.eu/legal-content/EN/TXT/?uri=COM%3A2017%3A7%3AFIN [Accessed 5 January 2021].

European Commission (2017b) 'Press Statement by Commissioner Věra Jourová, Mr. Lee Hyo-seong, Chairman of the Korea Communications Commission and Mr. Jeong Hyun-cheol, Vice President of the Korea Internet & Security Agency', 20 November, Statement/17/4739, Available from: https://ec.europa.eu/commission/presscorner/detail/en/STATEMENT_17_4739 [Accessed 5 January 2021].

Fischer-Hübner, S. (1997) 'Privacy at risk in the global information society', in J.J. Berleur and B.D. Whitehouse (eds) *An Ethical Global Information Society: Culture and Democracy Revisited*, Boston, MA: Springer, pp 261–73.

Fitzpatrick, M. (2015) 'Minority office report: Warning over software now being used by bosses that predicts if you're going to steal from the firm, have a nervous breakdown and even have an office affair', *MailOnline*, 17 February, Available from: https://www.dailymail.co.uk/news/article-2956766/Minority-office-report-Warning-software-used-bosses-predicts-going-steal-firm-nervous-breakdown-office-affair.html [Accessed 5 January 2021].

Gallagher, R. (2018) 'The untold story of Japan's secret spy agency', *The Intercept*, 19 May, Available from: https://theintercept.com/2018/05/19/japan-dfs-surveillance-agency/ [Accessed 5 January 2021].

Gascón Marcén, A. (2016) 'La nueva protección de datos personales: Una mirada a Japón desde Europa' [The new protection of personal data: A look at Japan from Europe], in C. Tirado Robles and F. Barberán Pelegrín (eds) *Los Derechos Individuales en el Ordenamiento Japonés* [Individual Rights in the Japanese System], Cizur Menor: Aranzadi, pp 129–48.

Gascón Marcén, A. (2020) 'Society 5.0: EU-Japanese Cooperation and the Opportunities and Challenges Posed by the Data Economy', ARI 11/2020, Real Instituto Elcano, Available from: http://www.realinstitutoelcano.org/wps/portal/rielcano_en/contenido?WCM_GLOBAL_CONTEXT=/elcano/elcano_in/zonas_in/ari11-2020-gascon-society-5-0-eu-japanese-cooperation-and-opportunities-and-challenges-posed-by-data-economy [Accessed 5 January 2021].

Greenleaf, G. (2015) 'Japan: Toward international standards – except for "big data"', *Privacy Laws & Business International Report*, 135: 12–14.

Greenleaf, G. (2018a) 'Japan: EU adequacy discounted', *Privacy Laws & Business International Report*, 155: 8–10.

Greenleaf, G. (2018b) 'Japan and Korea: Different paths to EU adequacy', *Privacy Laws & Business International Report*, 156: 9–11.

Greenleaf, G. and Park, W-i. (2014) 'South Korea's innovations in data privacy principles: Asian comparisons', *Computer Law & Security Review*, 30(5): 492–505.

Horowitz, M., Randles, M., and Ram, L. (2015) *Hackable at Any Height: Cybersecurity in the Aviation Industry*, London: London Cyber Security.

International Telecommunication Union (2018) 'Statistics [of Individuals Using the Internet from 2005–2019', Available from: https://www.itu.int/en/ITU-D/Statistics/Pages/stat/default.aspx [Accessed 5 January 2021].

Jee, C. (2019) 'Amazon's system for tracking its warehouse workers can automatically fire them', *MIT Technology Review*, 26 April. Available from: https://www.technologyreview.com/f/613434/amazons-system-for-tracking-its-warehouse-workers-can-automatically-fire-th em/?fbclid=IwAR0gwuhG4vYGgGcYA1NwQWNNCXuFyjdBIf SrXcotFyVZBAl1NWu_04SmK4I [Accessed 5 January 2021].

Kanetake, M. and de Vries, S. (2018) 'EU-Japan economic partnership agreement: Data protection in the era of digital trade and economy', *Renforce Blog*, 18 December, Available from: http://blog.renforce. eu/index.php/nl/2018/12/18/eu-japan-economic-partnership-agreement-data-protection-in-the-era-of-digital-trade-and-economy-2/ [Accessed 5 January 2021].

Kanetake, M. and Taylor, M. (2017) 'A right to be forgotten case before the Japanese Supreme Court', *Renforce Blog*, 7 February, Available from: http://blog.renforce.eu/index.php/en/2017/02/07/a-right-to-be-forgotten-case-before-the-japanese-supreme-court/ [Accessed 5 January 2021].

Kang, C. H. and Kim, S. H. (2020) 'Recent major amendments to three South Korean data privacy laws and their implications', *International Bar Association*, 11 June, Available from: https://www. ibanet.org/Article/NewDetail.aspx?ArticleUid=0D5FD702-179C-42A1-B37D-45D12F4556DA [Accessed 5 January 2021].

Kim, H., Kim, S. Y., and Joly, Y. (2018) 'South Korea: In the midst of a privacy reform centered on data sharing', *Human Genetics*, 137(8): 627–35.

Ko, H., Leitner, J., Kim, E., and Jung, J-G. (2016) 'Structure and enforcement of data privacy law in South Korea', Brussels Privacy Hub Working Papers, 2(7), Available from: https://brusselsprivacyhub. eu/publications/wp27.html [Accessed 5 January 2021].

Mallard, W. and Sieg, L. (2015) 'Japan pension system hacked, 1.25 million cases of personal data leaked', *Reuters*, 1 June, Available from: http://www.reuters.com/article/us-japan-pensions-attacks-idUSKBN0OH1OP20150601 [Accessed 5 January 2021].

Masao, H. (2012) 'Digital identity management and privacy in Japan', in J. Bus, M. Crompton, M. Hildebrandt, and G. Metakides (eds) *Digital Enlightenment Yearbook 2012*, Amsterdam: IOS Press, pp 97–108.

Mayer-Schoenberger, V. and Cukier, K. (2013) *Big Data: A Revolution that Will Transform How We Live, Work, and Think*, London: John Murray.

Meyer, D. (2018) 'South Korea's EU adequacy decision rests on new legislative proposals', International Association of Private People, 27 November, Available from: https://iapp.org/news/a/south-koreas-eu-adequacy-decision-rests-on-new-legislative-proposals/ [Accessed 5 January 2021].

Micheli, M., Lutz, C., and Büchi, M. (2018) 'Digital footprints: An emerging dimension of digital inequality', Journal of Information, Communication and Ethics in Society, 16(3): 242–51.

Miyashita, H. (2011a) 'The evolving concept of data privacy in Japanese law', International Data Privacy Law, 1(4): 229–38.

Miyashita, H. (2011b) 'Consumer agency takes charge of Japan's DP regime', Privacy Laws & Business International Report, 103: 2–3.

Miyashita, H. (2014) 'Japan appoints new independent commission for the supervision of ID numbers', Privacy Laws & Business International Report, 127: 10–12.

Miyashita, H. (2015) 'Japan amends its DP Act in light of Big Data and data transfers', Privacy Laws & Business International Report, 137: 8–11.

Murata, K. (2004) 'Is global information ethics possible? Opinions on the technologically-dependent society', Journal of Information, Communication and Ethics in Society, 2(5): 518–19.

My Number Promotion Office (2015) 'My number system: Basic outlines and future plans', [presentation slides], Available from: https://www.cao.go.jp/bangouseido/pdf/my_number_system.pdf [Accessed 5 January 2021].

Nakada, M. and Tamura, T. (2005) 'Japanese conceptions of privacy: An intercultural perspective', Ethics & Information Technology, 7(1): 27–36.

Orito, Y. and Murata, K. (2005) 'Privacy protection in Japan: Cultural influence on the universal value', Proceedings of ETHICOMP 2005, Linkoping: Linkoping University.

Orito, Y. and Murata, K. (2008) 'Rethinking the concept of information privacy: A Japanese perspective', Journal of Information, Communication and Ethics in Society, 6(3): 233–45.

Park, K.B., Ko, H.K., Chae S.H., and Kang, M. (2018) 'Korea's proposed overhaul of data protection laws', Privacy Laws & Business International Report, 156: 12–18.

PPC (2020) 'The Amendment Act of the Act on the Protection of Personal Information, etc. (Overview)', Available from: https://www.ppc.go.jp/files/pdf/overview_amended_act.pdf [Accessed 5 January 2021].

Shoniregun, C.A., Dube, K., and Mtenzi, F. (2010) Electronic Healthcare Information Security, Dordrecht: Springer Science & Business Media.

Tkach-Kawasaki, L.M. (2015) 'Japan', in *Freedom of the Net 2015*, Freedom House, pp 451–66. Available from: https://freedomhouse. org/sites/default/files/2020-02/Freedom_on_the_Net_2015_ complete_book.pdf [Accessed 5 January 2021].

Van Dijck, J. (2014) 'Datafication, dataism and dataveillance: Big data between scientific paradigm and ideology', *Surveillance & Society*, 12(2): 197–208.

Wall, A. (2018) 'GDPR matchup: South Korea's personal information protection act', International Association of Privacy Professionals, 8 January, Available from: https://iapp.org/news/a/gdpr-matchup-south-koreas-personal-information-protection-act/ [Accessed 5 January 2021].

Zuboff, S. (2019) *The Age of Surveillance Capitalism: The Fight for a Human Future at the New Frontier of Power*, London: Profile Books.

4

Phenomena and Phobia through *Pokémon* GO: An Analysis of the Reactions on the Augmented Reality Game in Japan

Deirdre Sneep

Introduction

It is a typical Sunday in Akihabara in May 2019. Just outside of the station's Electric Town exit, about 50 people gathered, closely packed (see Figure 1). The group included both men and women (although admittedly, more of the former) who ranged from middle school students to people in their fifties. Although there were small groups of people that appeared to know each other, most were on their own and did not interact with the others. The reason this group of strangers gathered in this particular spot was that they were meeting each another in a virtual space. Eyes glued to their screens, they were engaging in a world of augmented reality through the game *Pokémon GO*.

Many will still remember the sudden boom of players that this application brought into the streets in the summer of 2016. The application left, in many parts of the world, a real digital legacy. What is especially interesting for researchers who are interested in the 'homo ludens', however, is the way this smartphone game has been

Figure 1: Dozens of *Pokémon GO* players gathered in front of Akihabara station, Japan, in May 2019 (photo by Deirdre Sneep)

perceived and discussed. *Pokémon GO* seems to invite – as so many digital milestones do – a great emotional response. Although the game won – and still wins – great praise from enthusiasts, it has also drawn a wave of negative reactions, underlined with fear.

When it comes to the relationship between society and technology, if there is one emotion that continues throughout human history, it is fear. In this chapter, the phobic reactions to the phenomenon of *Pokémon GO* in Japan are analysed and used as a lens to shine new light on the issue of smartphone-related technophobia.

Background

Pokémon GO is an augmented reality smartphone gaming app developed by Niantic in co-operation with Nintendo and The Pokémon Company. Players' main objective is to walk around in physical locations to catch and train in-game imaginary animals (called Pokémon). When users bring their smartphones to specific places in real-world public locations such as parks, streets, or public buildings (for example, museums), their movements trigger specific in-game actions. The in-game map excludes private properties such as houses. As players walking around playing the game on their smartphones, their virtual avatars walk around within the game's virtual world as well. *Pokémon GO* uses the device's GPS and renders the virtual world for the game character over the phone's GPS data of the physical environment. For example, during a walk in a park, a player might 'encounter' a Pokémon and receive a notification on their screen. When opening the notification, the application would show a cartoon of the Pokémon, hovering on the screen with the view from the camera application as a backdrop. As the cartoon is 'anchored' in a specific location, the user will lose sight of the monster when turning the camera in another direction. The player will then typically engage in a 'fight' with the Pokémon, eventually either catching the in-game monster or losing it. All of these interactions use the image captured by the phone's camera as a backdrop, making the phone a 'glass' that reveals the hidden world of Pokémon. Because the in-game monsters, treasures, and other points of interests that spawn at certain locations – specific points within cities – are all projected on the screen over the camera's representation of the real world, the application creates a mix of the real and the virtual.

The Akihabara station exit was important to the players because it is a location packed with in-game items and rare Pokémon. Players are notified of these items and monsters through the in-game map, which uses an adapted version of the phone's GPS and is a representation of the real map of the player's environment. Players also often report their finds online on message boards. The game does not have a multi-player mode, but reporting finds online is still a very social experience for players. As Akihabara's Electric Town station exit is a well-known hub for in-game Pokémon and treasures, the station attracts large numbers of *Pokémon GO* players every day. The station exit is supposedly a good place to try one's luck at catching a rare monster. While there are ways and strategies to increase the chances to catch the rare finds, much of the game depends on luck, making every encounter a gamble in itself. The players who gather here are not only Japanese: Foreign tourists also arrive at the Electric Town exit, because there are some Pokémon that can only be caught in Japan. Although the game itself might not be the sole reason for foreign visitors to travel to the country, it is undeniable that the Pokémon lore – the games, animation, and merchandise – has a large fanbase in many countries around the world, and might be one of the reasons for many people to visit Japan.

Akihabara has been a place for electronic gaming since the late 1980s. Before the 1990s, the neighbourhood's stores mainly sold household appliances and supplies for electric circuits. Driven by the progress of information technology which made personal computers part of the in-house electronics of the average family, Akihabara, previously known only for household electronic appliances, became a place for *otaku* (nerds) and gamers in the 1990s and 2000s (Morikawa, 2003). It has been pointed out, however, that the 'electric town' has become much more touristy since the 2000s and has thus transformed from an exclusive hub for the Japanese *otaku* and gamers to a place for foreign tourists to experience Japan's popular culture, in particular the culture of video games, manga, and anime (Galbraith, 2006; see also Chapter 2 for more about the location). *Pokémon GO* has given this Tokyo neighbourhood another brand new meaning in terms of digital play. The addition of a virtual layer on top of the location (albeit one only possible with the help of a digital device) transforms a place where the only physical experience of play is shopping for video games and merchandise into an area of virtual-reality play, adding a new physical experience to gamers. While Akihabara already was a hot spot as a location of 'play' because of the nature of the shopping centre and historic connection with video games and electronics, the fact that it has become a hotspot for 'virtual reality play' has reinforced and amplified the meaning of this location in Tokyo.

But Akihabara is not the only neighbourhood that has taken on new meaning after the smartphone game fad swept the country. Player hubs like these can be found scattered throughout Japanese cities. Tokyo alone already has dozens of these Pokémon hotspots. Flocks of gamers visit them to gather special (and sometimes limited edition) in-game items, defeat virtual opponents, and catch and train Pokémon characters. An equally (if not more) popular spot than Akihabara's station exit can be found in Ueno Park, not too far from Akihabara. In 2016, Ueno Park was such a popular point of interest for players that the hundreds of smartphone users who instantly downloaded the game after its launch caused stampedes in their search for Pokémon in the park (Baseel, 2016). While there are fewer active players nowadays, a significant number of them can still be seen playing around these places. Especially on days where there are special events in-game, large crowds of players gather, and surrounding stores see an opportunity to stack up on portable chargers to sell to desperate smartphone users who are low on battery and to provide benches for the augmented reality gamers to sit on (Blaster, 2018).

A few months after its release on 22 July 2016, *Pokémon GO* was a worldwide phenomenon, and popular in-game locations all over the world attracted sometimes hundreds of people at the same time (Baseel, 2016; Ihme, 2016; Sanda, 2016). The hype gradually reduced down in many countries, and the game was labelled a 'fad' that had enjoyed a short-lived period of popularity. In the end, it died down once the summer months had passed. In Germany, for example, the smartphone game was reported to have 'died' a few months after its launch and urban environment researchers argued that after its 'death' there were no 'long-term effects' of the game. In Germany, the game has not changed pedestrian behaviours or traffic patterns. There was, for example, no need to increase the number of benches (for players to sit on) or other street furniture that would change the look of the city (Andone et al, 2017). However, as with so many trends related to mobile phones and digital culture in general (Hjorth and Chan, 2009a; McLelland, 2013), the success of *Pokémon GO* proved to be locally specific. Unlike in European countries, in Japan *Pokémon GO* is far from extinct as of 2019, and players of the virtual reality game can still be seen throughout the cities. As a matter of fact, a whole new 'boom' in player activity was reported in autumn 2018, two years after the game's launch (Fukatsu, 2018).

To study the case of *Pokémon GO* in Japan, we should actually begin from before the game was officially launched. *Pokémon GO* was launched two weeks later in Japan than in the US, where the smartphone game was developed. This delay in release resulted in people downloading the game from illegal (Chinese) websites so that they did not have to wait for the official launch. The delay also resulted in increased anticipation in Japan. In the US, the game was an almost immediate success, and it did not take long for news stories from the US about the augmented reality game to be spread worldwide. Spurred by some of the more catastrophic stories, the National Center of Incident Readiness and Strategy for Cybersecurity (NISC) in Japan issued a nine-point pamphlet through the online platforms Twitter and LINE, with warnings against the risks of playing *Pokémon GO*. The pamphlet, published just a few days before the launch of the game in Japan, included a warning not to enter unsafe places, to be careful of overheating devices, to be cautious of scammers, and to definitely never use one's phone while walking (NISC, 2016).

The pamphlet, titled 'Requests for all Pokémon Trainers' was issued online through social media before the launch of *Pokémon GO* in Japan by the NISC. Illustrated with cartoons, the pamphlet alerts *Pokémon GO* players of nine important points that they should be mindful of when playing the augmented reality game. Point Seven includes an illustration of a dangerous Westerner in a cowboy hat and says 'in other countries, people have already been run over by cars, have fallen into water, been bitten by snakes, and robbed by thieves all while playing *Pokémon GO* while walking. Please stay out of places that look dangerous.' Point Nine reads:

> A lot of accidents are happening because people use their phones while walking. People have been injured both on station platforms due to trains and on the street by cars. Smartphone [use while] walking is extremely dangerous. When your phone buzzes because of the game, stop walking and look around you. Of course, it goes without saying that playing on a bike is also not acceptable. (NISC, 2016)

Technophobia and the smartphone

While Japan is often portrayed as a country that has a high affinity with technology (Morley and Robins, 1995), it has in fact been suspicious of new digital technologies, showing an interesting stance towards the digitalisation of society. So far, Japan has undergone several 'cycles' of moral panic-like outbursts regarding digital technologies, starting in the early 2000s with the fear of '*deai kei*' (net forums) and the various problems these were rumoured to bring, such as '*netto ijime*' (cyberbullying), (child) prostitution, and (mass) '*netto jisatsu*' (suicide plots) (Joseph et al, 2003). Suspicion towards information technology continued into the mobile phone era, with fears that devices had a plethora of unwanted effects (especially on teenagers) and were associated with mental illnesses (see, for example, Takao et al, 2009; Ikeda and Nakamura, 2014). In the early 2000s, mobile phone 'anxiety' discourse grew so large that, according to Takahashi (2011), it did not reflect the reality of mobile phone use anymore. According to her, the prominent position of teenagers in this discourse suggested that its main function was to emphasise a difference between generations, and that it was more reflective of the hopes and fears of older generations – in particular the fear of the collapse of the family, schools, and society (Takahashi, 2011). In today's age, Takahashi explains, social media applications are actually a core aspect of self-identity among young Japanese people, something that is difficult for older generations to understand, less alone to exert control over, as the mobile phone is a personal device that does not invite shared usage (Takahashi, 2011; 2016). This indicates a difference in perspective on mobile technology between the older and younger generations. While the younger people experience mobile phone technology as a central part of their lives, both for work and education as well as for leisure and play, the older generation seems to be inclined to reject this perspective, and instead sees mobile technologies as primarily functional – that is, only for work but not for leisure or play.

However, this only explains the discourse of fear of mobile devices with regard to teenagers and children. Another more general explanation likely plays a part as well. The fearful tendency in discourse regarding the spread of new technologies, after all, is not unusual nor unique to Japan. New technologies are always accompanied by a certain sense of risk, fear of which is sometimes referred to as 'technophobia' (Dinello, 2005). This sense of risk can spread throughout society in the form of irrational fears and concerns, such as the ones we see that are attached to mobile phone technologies. Often, this is expressed through the upsetting idea that technology is infiltrating our bodies (Lupton, 1995), reflecting the fear that technology is consuming *us* instead of the other way around (Silverstone and Hirsch, 1992). In the case of mobile phones, 'smartphone-phobia' tends to centre on the fear of the device 'hi-jacking' the nervous system – reminiscent of the ways the Internet, and desktop computers and video game devices before it, were imagined to affect users' brains (Marshall, 1997; Brosnan, 1998). Such phobia portrays the smartphone as numbing our senses and stupefying our brains, depicting mobile-phone use as highly addictive and its use as pathological. The inclination to associate mobile-phone use with bodily dysfunctions and addictions seems to be common across cultures, and we see a high number of studies in Japan regarding this topic. This would explain why the discourse on mobile phone-related issues in Japan in the 2000s focused so heavily on problems related to the body, especially mental health.

In the phobias that accompany the smartphone, the intimate connection of the user and the device is always emphasised. According to some, this is because the mobile phone, as a device that is constantly close to the body and has a particularly close relation to our senses of seeing, touching, and hearing, can be seen as some sort of 'extension' of ourselves (see, for example, Grosz, 1994; Fortunati, 2003a, 2003b; Fortunati et al, 2003; Turkle, 2005; see also Chapter 5 for the materiality and corporeality of media technologies). As the body is in constant interaction with the phone, the body eventually experiences the device as a part of itself. This is not true for the mobile phone alone: Neurological research reveals that tools that are used constantly and are worn close to or on the body, for example glasses and watches, remap the brain to 'incorporate' them into a person's body schema (Clark, 2007; 2008; Hanna and Maise, 2009). On top of that, the relationship between a person and their mobile phone is also *emotionally* very intimate, as it is a tool that is used primarily for social connection. This in turn might also influence the way people hold the device, often trying to keep it closer to their bodies (Vincent et al, 2005). In a sense, the mobile phone user can be seen as a 'cyborg', a human who has technology incorporated (Haraway, 1991). Only unlike the image of the transcendent cyborg, the smartphone user – especially in the case of those using the device while walking – is often portrayed as an unhealthy addict. Lupton (1995) previously argued that the bodies of those who are branded by society as Internet or computer addicts tend to be portrayed in deformed ways. Although the smartphone user is not usually portrayed as deformed, there are most definitely public concerns on how smartphone use leads to health issues such as neck issues, eye problems, and back pain (Kim et al, 2013; Maniwa et al, 2013; Kim, 2015; Xie et al, 2016). In 2015, there was even a short online 'panic' going viral on various social media on how excessive smartphone use would lead to finger deformations, including images of so-called smartphone pinkies. For example, on 4 March 2015 mobile phone provider DoCoMo tweeted out an image to raise awareness of the 'smartphone pinky': a deformation of the finger that is allegedly the result of inflammation due to overuse of the smartphone (DoCoMo Official Support, 2015). In the picture, we see a hand using a smartphone and then a close up of the hand, showing an indent in the pinky finger and a slight deformation in the bone structure, right at the place on where one would rest one's phone. In order to prevent the 'smartphone pinky', the provider advises users to take breaks and switch hands when using a phone.

A hand surgeon later diagnosed the person's hands in the pictures that went viral with a condition unrelated to smartphone use, and stressed that 'smartphone pinky' is very unlikely to occur in an average person's hand (Ossola, 2016). In this particular case, the surgeon suggested that the person had Dupuytren's contracture, a condition where tissue in the fingers gradually stiffens over time, causing deformations such as in the pictures that were posted. Although the cause of the condition is unknown, it has been labelled the 'Viking disease' because of its prevalence among northern European men (Hart and Cooper, 2005).

From the early to mid-2010s, a new phobia surrounding digital technologies arose, this time centred on the discussion of smartphone use while walking, or '*aruki-sumaho*' as it is called in Japanese (see Fuseya, 2015; Masuda and Haga, 2015; Nishidate et al, 2016; Obara et al, 2016; Sekiguchi and Shun, 2016). Despite the pervasive fear, statistics show that there were fewer than ten individuals who got seriously injured due to smartphone use while walking in a time period of four years (2012–16) in all of Tokyo (Tokyo Fire Department, 2017), and thus the practice can hardly be considered dangerous. Nevertheless, it has been discussed so extensively that it invokes thoughts of a 'moral panic' (Cohen, 1972; see also Sneep, 2019). Not only have there been numerous campaigns to raise awareness of the phenomenon and calling for people to stop using their phones while walking, but there have also been several studies published on the topic. Smartphone use while walking has been called a 'social issue' (Sasao et al, 2016), and a 'dangerous side-effect of smartphone addiction' (Fuseya, 2015: 55). Some scholars pointed out that smartphone use while walking seemed to be particularly common among people who '*toku ni netto izon keikō no takai mono*' (have a tendency to be Internet addicts) (Nishidate et al, 2016: 109), again invoking images of mental health issues. Although there are other addictions common among people (such as smoking or excessive consumption of caffeine), smartphone addiction seems to be among the least socially accepted.

The *Pokémon* GO phenomenon

The concerns regarding *aruki-sumaho*, as well as the underlying phobia regarding addictive technologies (in particular video games), became amplified when *Pokémon GO* was launched in 2016. About a month after the release of the game, Japanese newspapers started to report on country-wide accidents and incidents caused by the new application. The *Sankei Shinbun* reported 727 cases of traffic violations related to the game, of which 15 resulted in personal injuries (*Sankei Shinbun*, 2016). The Tokyo Metropolitan Police Department reportedly took 553 minors into protective custody during the course of three months after the launch of the game (Satō, 2016a). Some of them were arrested because they played the game in dangerous places in the middle of the night; others were spotted littering in public areas while playing the game. Several minors were also arrested for shoplifting related to the smartphone application – some youngsters tried to steal fully charged mobile phone external batteries from shops when the game had exhausted their smartphone's energy supply (Satō, 2016a). Numerous similar incidents were also reported by the Japanese police, some leading to concerns of moral decline as the smartphone game led youngsters to behave inappropriately in public places.

From mid-2016 until mid-2019, negligence due to playing games led to the deaths of six people. Most of these cases were due to people operating vehicles while playing the game, despite the fact that the game has built-in protection that halts in-game activity when the player is found to be moving too fast (that is, faster than a brisk walking pace). To play while driving, one must somehow bypass this mechanism (although one can open the game's interface at any time without actually playing it, which can lead to distraction as well). A 39-year-old woman in Kyoto, for example, got stuck with her bike in the crane of a truck while it was waiting for a red light. The driver, unaware of what happened because he was reportedly distracted by the smartphone game, resumed driving without paying attention to his crane, leading to the unfortunate death of the woman (Satō, 2016b). Several other cases of reckless drivers playing the game were reported throughout 2016, with three more deaths that same year (*Asahi Shimbun*, 2016a; *Tokushima Shimbun,* 2016; *Sankei News*, 2017). But the most heartbreaking case was probably that of a nine-year-old boy in Aichi Prefecture who died after a truck driver drove into him and a group of other schoolchildren while playing *Pokémon GO* behind the wheel (*Shūkan Jyosei*, 2016). The most recent case of a death related to *Pokémon GO* was in 2018, when an 85-year-old woman, also in Aichi prefecture, was hit by a car driven by a woman in her forties who was (reportedly) distracted by reading the game's safety notification (Sanspo, 2018).

Another widely reported case that went viral on social media was that of a bus driver in Okayama in October 2016, although it did not involve any accidents. One of the passengers of a sightseeing bus operated by a popular tourism company uploaded a video of the driver playing *Pokémon GO* while driving, thereby putting at risk the lives of dozens of passengers on board ('*Asahi Shimbun*, 2016b; Kaminuma, 2016). The full video has so far been viewed more than 115,000 times on YouTube and includes comments such as '*hidoi yo no naka ni natta ne*' (what a cruel world this has become [to put lives at stake like this])', and '*kōyū ningen ni ha hajime ni shinde hoshii ne heiwa ni kurasu hitobito no tame ni mo*' (these kinds of people should hurry up and die to leave the rest of us in peace). Reactions to the video came almost exclusively from people who condemned the driver's actions, indicating a strong moral judgement of such reckless behaviour and an underlying fear of moral decline. Some of the commenters seem to blame the video game, saying it is the worst (*saiaku*) and implying the game is causing problems in society.

Although the individual cases are heartbreaking, the overall number is only a small percentage of overall traffic injuries and is in fact negligible when it comes to traffic accidents in Japan. One study, conducted during and after the peak of the *Pokémon GO* fad in Japan in 2016, argued: 'despite public concerns, [...] the effect of *Pokémon GO* on fatal traffic injuries may be negligible' (Ono et al,

2018: 487).[1] Apart from the few cases of people playing the game while driving, most incidents were much less dangerous. Most cases of *Pokémon GO* players getting into trouble involved people getting lightly injured by falling, tripping, or bumping into bicycles. Some of the reported accidents had little to do with the game at all, let alone with mobile phones in general. For example, *The Japan Times* reported that a 22-year-old woman was robbed while playing the game and not paying attention to her belongings. They also reported how a 24-year-old Brazilian tourist was found lost and wandering around looking for special Pokémon (*The Japan Times*, 2016). Although the cases are related to the game, distraction can hardly be ascribed solely to playing *Pokémon GO*.

On the other hand, while the media focused on the incidents and traffic accidents, players reported that the game was very enjoyable and it was even believed to have health benefits because of the increased physical activity that it encourages. Players stated that one of the things that changed most after they started playing the game was that they started to walk more (Senoo, 2016; Nigg et al, 2017; Nishiwaki and Matsumoto, 2018). Forty-three per cent of the players in Japan reported after playing for a month that they enjoyed themselves a lot (*totemo tanoshii*), and 34 per cent noticed that they were walking more. And although some have argued that playing the game would (intentionally or unintentionally) increase social isolation or social distance (Humphreys, 2017), 18 per cent of players actually said that they felt the game made them interact more with family or friends (Senoo, 2016). Despite these reported positive effects, popular media has tended to focus on sensational stories reported by the police or featuring victims of reckless drivers who were playing the game. These stories evoke more emotional responses and are more in line with the general association of smartphone use with moral decline.

Technophobia revisited

The Japanese reaction towards the game is particularly interesting for two reasons. First of all, smartphone games are a big part of what can be considered Japan's mobile phone culture, and *Pokémon GO* has enjoyed tremendous popularity. When it was finally released in Japan, a study of 1,949 smartphone users showed that more than one in three had downloaded and played the game within four days after its release, six times more than in the US, where (only) 5.6 per cent of smartphone users downloaded the game within four days (Evangelho, 2016; Senoo, 2016).[2] The reason *Pokémon GO* enjoyed so much more popularity in Japan than in the US could be the fact that the Pokémon franchise, which originated in Japan, might be culturally more tailored to the Japanese gaming audience, as a big part of the franchise is built around 'cute' (*kawaii*) characters, resonating with a love for cute characters among the Japanese. Anne Allison (2003) explains that when the franchise was exported from Japan to the US, a part of the cuteness that was attributed to the characters was actually purposefully removed to make it seem more appealing to a wider US market. For example, they made the characters' colours brighter and less 'soft'. In addition, they gave Pikachu's owner, a teenage boy, more prominence in the game than the cute Pikachu (a small, yellow, mouse-like Pokémon) (Allison, 2003). On the other hand, some have argued that the Pokémon franchise is relatively culturally odourless despite these minor amendments in the franchise (Iwabuchi, 2004).

Another, perhaps more likely explanation would be the fact that smartphone games are generally much more popular in Asia (Jin, 2017), especially in Japan (Hjorth and Richardson, 2017). Japan has a long history not just of mobile phone use, but also of mobile gaming devices, as evidenced by the popularity of devices such as Sony's PlayStation Portable and Nintendo's Game Boy (Hjorth, 2016; Hjorth and Richardson, 2017). Due to the history of gaming culture in which portability is a key element, the mobile phone is a popular medium not only because of a technological, but also socio-cultural affinity (Hjorth, 2016; Hjorth and Chan, 2009b). The popularity of smartphone games can be seen especially in areas of mobility such as public transportation, one of the spaces where smartphone gaming is ubiquitous. Mobile phone game developers regularly buy up large amounts of advertisement space in train stations to promote new content. This further reinforces the idea that smartphone gaming is a central part of daily life in Japan.

If smartphone games are so widespread and well-enjoyed in Japan, how come that the public discourse in popular media in Japan, as well as a large part of the scholarly discourse, has been preoccupied with such a trivial part of the *Pokémon GO* phenomenon, which emphasises the parts of the discourse that stir technophobic reactions the most? The boundaries between machine and human minds have become blurred in the digital age, and a general discourse seems to have arisen that portrays this as problematic. In the same discourse, more user behaviour is attributed to the machine than to the human mind. This is visible in the discussion of *aruki-sumaho*, and even more so in the case of *Pokémon GO*. The user is seen as completely driven by the demands of the game and the smartphone, so much so that the smartphone is depicted as having 'taken over' the nervous system of the user, along with their nervous system, brain and ability to function in a socially acceptable way.

In 1984, around the start of the golden era of the Japanese bubble economy and when the information technology revolution started to have a significant impact on people's daily lives, Kogawa Tetsuo wrote an article in the *Canadian Journal of Political and Social Theory* about the digitalisation of music and the negative psychological effect this would have on society, which echoes previous historical statements about machinery, while placing it in the temporal zeitgeist of the 1980s:

> As far as the present Japanese collectivity is concerned, it is electronic and very temporal, rather than a conventional, continuous collectivity based on language, race, religion, region or taste. A more vivid example of electronic collectivity is the 'Walkman', a tiny cassette player with a headphone. 'Walkman' is already an internationally popular commodity of SONY, but it has a special meaning in Japanese society. This device is for personal use only, so that the users are isolated from each other even if they listen to the same music source. They are united in the way of marionettes. 'Walkman' users form a collectivity of marionettes, maintaining some kind of individualism of the user. [...] Most users of 'Walkman' in Japan are listening to FM radio, which is awfully monotonous and is not free from the control of the government and the culture industries. (Kogawa, 1984: 18)

Here Kogawa compares Walkman users to marionettes, implying that the technology of the Walkman is somehow taking over their brains. A comparison between 'Walkman marionettes' and 'Smartphone zombies' springs to mind. Although Kogawa is pointing fingers at government control, nowadays other factors are being blamed – depending on the era of digitalisation, the scapegoats tend to shift as well. But Kogawa's statement also shows that the idea that the technology is infiltrating the brain is persistent and not unique to the 'smartphone era'. In Japan, the technophobia that eventually resulted in the moral panic-like concern around *Pokémon GO* is the direct result of a fear of information technology that started decades ago and transformed through the 1980s, 1990s, and early 2000s into its current form.

As a matter of fact, it is not even the first time that the Pokémon franchise has been linked to neurological concern. In December 1997, hundreds of children in Japan experienced seizures that were induced by a strobe light effect in an episode of the Pokémon animated series that was broadcast on television. In response to this incident, also labelled as Pokémon Shock (*Pokemon shokku*), the series was halted until the cause of the incident was found. In his analysis of Pokémon Shock, Lamarre (2018) shows that the reactions to the incident largely repeated an earlier technophobic image of the television as a brain intoxicant. While the cause of the incident was relatively quickly located and 'fixed', a mild hysteria followed, with thousands of parents reporting to medical authorities that their child might be experiencing the same problem, even though the children showed minor to no symptoms (Radford, 2001). It was even argued by some that it was not just that one episode of Pokémon on television that caused seizures, but that the video games had had the same effect on some people as well, and that Nintendo had purposefully kept this aspect of their franchise secret (Lamarre, 2018).

This incident, although with a different medium as its vessel, has some striking similarities to the case of *Pokémon GO*. Just as with *Pokémon GO*, the phobic reactions are augmented by the concern for children's health. Further, as is more obvious in the case of Pokémon Shock, there is an underlying assumption that there is somehow a close link between neurological disorder and the electronic screen. As a result, the discourse of machines infiltrating the body only gets repeated and reinforced. Lamarre wrote:

> [N]o one takes seriously proposals to abandon televisions, computers, or cell phones [...]. We opt instead for regulation, which tends to rehystericize the understanding of power, attributing quasi-magical agency to images, television sets [...] while producing an understanding of the human body as inherently passive, frail, precarious, and under siege, and hence in need of regulatory governance based on studies of neurological development. (Lamarre, 2018: 86)

Conclusion

I have shown in this chapter that '*Pokémon GO* phobia', as it has been discussed by the media and scholars in the years since its release, is actually part of a series of techno-phobias in Japan, which have followed Japan's own unique relationship with information technologies over the decades. Starting in the 1980s, alongside the information technology revolution, Japan's relationship with the Internet and digital media has included a fear of digital technologies similar to what has also been seen in other countries. We see an upheaval especially after the 2000s, when the 'moral panic' surrounding online message boards, cyberbullying, Internet suicide pacts, and, later, the fear of the effects of mobile phone technology on mental and physical health (especially among youngsters) lead to much widely discussed research and many popular articles on the topic. Nonetheless, by the end of the 2010s, smartphones had taken a prominent place in Japanese society and smartphone gaming had become one of the most lucrative businesses. While smartphones were initially tools of work – their technological history is generally thought to have started with the development of Personal Digital Assistants (PDAs) in the 1980s and 1990s – they have long since transgressed the border between work and play. In fact, with the rise of the smartphone gaming industry and social network applications, smartphones have transformed into tools for evasion of work. A shift like this is also generational; while older generations use these devices as tools for work, the younger generation increasingly sees them as tools for play. This is especially true among the youngest generation, which will see the smartphone primarily as a tool for leisure. *Pokémon GO*, with thousands of players (especially younger ones) actively visiting locations in real life, is thus far one of the most visible smartphone games. The presence of the smartphone game in places that have previously not been used for gaming or play is noticeable and these changes in pedestrian behaviour are a significant change in the urban environment. Not only is the device shifting meaning from work to play, but it has also added an element of 'play' to physical space in the city. By doing so, it has immediately become a widely discussed topic, prone to accusations from older generations.

Technophobia has a long history and experiences episodes of revival whenever a new device or medium is introduced and becomes popular (Briggs and Burke, 2014). In the nineteenth century, people claimed machinery was compromising the human ability to live a good life. Technology would result in a life that is too 'mechanical' instead of one that is 'heroic' or 'devotional', according to the dominant ideas of the time about what it meant to live a meaningful life (Briggs and Burke, 2014: 53). Episodes of technophobia have resurged over the decades, especially as we entered the information age. Although there are common global trends when it comes to the spread of phobias in the digital age, they are also locally bound, as the case of smartphone use and *Pokémon GO* has also shown.

As a final note, although *Pokémon GO* has received much negative attention, the actual game and its players do not seem to be affected much by the negativity in popular media. Whenever the game receives new patches or updates, Japanese players can be seen swarming to the locations that have been 'updated'. Although *Pokémon GO* is the first augmented reality game to build such a large player base, it paves the way for a new kind of 'play' – one where the digital is increasingly interacting with physical place and location, potentially changing the experience of public space for future generations and conflicting with the traditional boundaries between work and play from previous generations.

Notes

[1] To extend the comparison, in 2016, 58 people needed medical attention after getting into accidents due to careless mobile phone use (in general), out of a total of 131,925 people needed medical help after getting into traffic accidents (Tokyo Fire Department, 2017; 2018). This means the percentage of mobile phone-related accidents was 0.04 per cent.

[2] However, it should be stated that the smartphone diffusion in Japan is lower than in the US (this was still the case as of 2016), and the demographic that owns a smartphone in Japan is generally younger on average, usually in their 20s and 30s. On average, *Pokémon GO* attracts a younger audience. Thus, seeing the age difference among smartphone users, this number is not entirely comparable.

References

Allison, A. (2003) 'Portable monsters and commodity cuteness: Pokémon as Japan's new global power', *Postcolonial Studies*, 6(3): 381–95.

Andone, I., Blaszkiewicz, K., Böhmer, M., and Markowetz, A. (2017) 'Impact of location-based games on phone usage and movement', *Proceedings of the 19th International Conference on Human-Computer Interaction with Mobile Devices and Services*, pp 1–8, Available from: https://doi.org/10.1145/3098279.3122145 [Accessed 4 January 2021].

Asahi Shimbun (2016a) 'Untenchū ni Poke GO, mata shibou jiko – wakimi de jyosei haneta utagai' [Again accident due to Pokemon Go while driving – suspect to have hit a woman while looking aside], 26 August, Available from: https://www.asahi.com/articles/ASJ8V351VJ8VOIPE004.html [Accessed 4 January 2021].

Asahi Shimbun (2016b) 'Kankō basu untenchū ni Pokemon GO kaisha "arumajikikoto"' [Playing Pokemon Go while driving a tourist bus – 'unfit to work' says company], 4 November, Available from: https://www.asahi.com/articles/ASJC441JSJC4PPZB006.html [Accessed 4 January 2021].

Baseel, C. (2016) 'Mini stampedes in Tokyo's Ueno Park as it becomes a nest for rare Pokémon GO character [video]', *Sora News 24*, 11 August, Available from: https://soranews24.com/2016/08/11/mini-stampedes-in-tokyos-ueno-park-as-it-becomes-a-nest-for-rare-pokemon-go-character [Accessed 4 January 2021].

Blaster, M. (2018) 'Tokyo's Shinjuku Station's west side: Quite possibly best spot in Japan for Pokémon GO', *Sora News 24*, 30 August, Available from: https://soranews24.com/2018/08/30/tokyos-shinjuku-stations-west-side-quite-possibly-best-spot-in-japan-for-pokemon-go/ [Accessed 4 January 2021].

Briggs, A. and Burke, P. (2014) *A Social History of the Media: From Gutenberg to the Internet*, Cambridge: Polity Press.

Brosnan, M.J. (1998) *Technophobia: The Psychological Impact of Information Technology*, London: Routledge.

Clark, A. (2007) 'Re-inventing ourselves: The plasticity of embodiment, sensing and mind', *Journal of Medicine and Philosophy*, 32(3): 163–82.

Clark, A. (2008) *Supersizing the Mind: Embodiment, Action and Cognitive Extension*, New York: Oxford University Press.

Cohen, S. (1972) *Folk Devils and Moral Panics*, London: MacGibbon & Kee.

Dinello, D. (2005) *Technophobia!: Science Fiction Visions of Posthuman Technology*, Austin: University of Texas Press.

DoCoMo Official Support (2015) '"Yubi ga henkei!?" Sumaho no mochikata ni yotte ha…' ['Finger Deformations!?' holding your smartphone may have effect on…], 4 March, Available from: https://twitter.com/docomo_cs/status/573332159911612416 [Accessed 4 January 2021].

Evangelho, J. (2016) '"Pokémon GO" is about to surpass Twitter in daily active users on Android', *Forbes*, 10 July, Available from: https://www.forbes.com/sites/jasonevangelho/2016/07/10/pokemon-go-about-to-surpass-twitter-in-daily-active-users/#1d9ec19b5d3e [Accessed 4 January 2021].

Fortunati, L. (2003a) *Mediating the Human Body: Technology, Communication, and Fashion*, London: Routledge.

Fortunati, L. (2003b) 'The human body: Natural and artificial technology', in J.E. Katz (ed) *Machines that Become Us: The Social Context of Personal Communication Technology*, Piscataway, NJ: Transaction Publishers, pp 71–87.

Fortunati, L., Katz, J.E., and Riccini, R. (2003) *Mediating the Human Body: Technology, Communication, and Rashion*, Mahwah, NJ: Lawrence Erlbaum.

Fukatsu, A. (2018) 'Naze ima "Pokemon GO" ga sai buumu? kaihatsusha ga kataru "shōhi sarenai shikake" to ha' [Why is there a new 'Pokemon Go' boom? Pokemon Go is an 'endless consumption' according to the developers], *Oricon News*, 13 November, Available from: https://www.oricon.co.jp/special/52082/ [Accessed 4 January 2021].

Fuseya, S. (2015) 'Actual condition on using smartphones for female students and danger of texting while walking', *Abstracts of Annual Congress of the Japan Society of Home Economics*, 67: 55.

Galbraith, P.W. (2006) 'Akihabara: conditioning a public "otaku" image', *Mechademia*, 5: 210–30.

Grosz, E.A. (1994) *Volatile Bodies: Toward a Corporeal Feminism*, Bloomington: Indiana University Press.

Hanna, R. and Maise, M. (2009) *Embodied Minds in Action*, New York: Oxford University Press.

Haraway, D. (1991) 'A cyborg manifesto: science, technology, and socialist-feminism in the late twentieth century', in D. Haraway (ed) *Simians, Cyborgs and Women: The Reinvention of Nature*, New York: Routledge, pp 149–81.

Hart, M.G. and Cooper, G. (2005) 'Clinical associations of Dupuytren's disease', *Postgraduate Medical Journal*, 81(957): 425–8.

Hjorth, L. (2016) 'Games of being mobile: The unruly rise of mobile gaming in Japan', in D.Y. Jin (ed) *Mobile Gaming in Asia: Politics, Culture and Emerging Technologies*, New York: Springer, pp 21–34.

Hjorth, L. and Chan, D. (2009a) 'Locating the game: Gaming cultures in/and the Asia-Pacific', in L. Hjorth and D. Chan (eds) *Gaming Cultures and Place in Asia-Pacific*, London: Routledge, pp 1–18.

Hjorth, L. and Chan, D. (eds) (2009b) *Gaming Cultures and Place in Asia-Pacific*, London: Routledge.

Hjorth, L. and Richardson, I. (2017) 'Games of being mobile: The unruly rise of mobile gaming in Japan', in D.Y. Jin (ed) *Mobile Gaming in Asia: Politics, Culture and Emerging Technologies*, New York: Springer, pp 21–33.

Humphreys, L. (2017) 'Involvement shield or social catalyst: thoughts on sociospatial practice of Pokémon GO', *Mobile Media and Communication*, 5(1): 15–19.

Ihme, L. (2016) 'Girardet-Brücke in Düsseldorf: Pokémon-Jäger wünschen mehr Platz' [Girardet Bridge in Düsseldorf: Pokémon hunters want more space], *RP Online*, 25 July, Available from: https://rp-online.de/nrw/staedte/duesseldorf/girardet-bruecke-in-duesseldorf-pokemon-jaeger-wuenschen-mehr-platz_aid-18105827 [Accessed 4 January 2021].

Ikeda, K. and Nakamura, K. (2014) 'Association between mobile phone use and depressed mood in Japanese adolescents: A cross-sectional study', *Environmental Health and Preventive Medicine*, 19(3): 187–93.

Iwabuchi, K. (2004) 'How "Japanese" is Pokémon', in J. Tobin (ed) *Pikachu's Global Adventure: The Rise and Fall of Pokémon*, Durham, NC: Duke University Press, pp 53–79.

The Japan Times (2016) '"Pokemon Go" craze leads to spate of accidents, traffic offenses across Japan', 25 June, Available from: https://www.japantimes.co.jp/news/2016/07/25/national/pokemon-go-craze-leads-spate-accidents-traffic-offenses-across-japan/ [Accessed 4 January 2021].

Jin, D.Y. (ed) (2017) *Mobile Gaming in Asia: Politics, Culture and Emerging Technologies*, Dordrecht: Springer.

Joseph, T., Holden, M., and Tsuruki, T. (2003) 'Deai-kei: Japan's new culture of encounter', in N. Gottlieb and M. McLelland (eds) *Japanese Cybercultures*, London: Routledge, pp 34–49.

Kaminuma, Y. (2016) 'Basu untenshu ga unten shinagara pokemon GO' ['Playing Pokemon GO while driving a bus'], YouTube [video], 24 October, Available from: https://www.youtube.com/watch?v=FYpkJ8aYMn8 [Accessed 4 January 2021].

Kim, M-S. (2015) 'Influence of neck pain on cervical movement in the sagittal plane during smartphone use', *Journal of Physical Therapy Science*, 27(1): 15–17.

Kim, Y.G., Kang, M.H., Kim, J.W., Jang, J.H., and Oh, J.S. (2013) 'Influence of the duration of smartphone usage on flexion angles of the cervical and lumbar spine and on reposition error in the cervical spine', *Physical Therapy Korea*, 20(1): 10–17.

Kogawa, T. (1984) 'Beyond electronic individualism', *Canadian Journal of Political and Social Theory*, 8(3): 15–20.

Lamarre, T. (2018) *The Anime Ecology: a Genealogy of Television, Animation, and Game Media*, Minneapolis: University of Minnesota Press.

Lupton, D. (1995) 'The embodied computer/user Runner', *Body & Society*, 1(3–4): 97–112.

McLelland, M.J. (2013) 'Socio-cultural aspects of mobile communication technologies in Asia and the Pacific: a discussion of the recent literature', in G. Goggin (ed) *Mobile Phone Cultures*, London: Routledge, pp 124–34.

Maniwa, H., Kotani, K., Suzuki, S., and Asao, T. (2013) 'Changes in posture of the upper extremity through the use of various sizes of tablets and characters', in S. Yamamoto (ed) *Human Interface and the Management of Information: Information and Interaction Design*, Berlin and Heidelberg: Springer-Verlag, Available from: https://doi.org/10.1007/978-3-642-39209-2_11 [Accessed 4 January 2021].

Marshall, P.D. (1997) 'Technophobia: video games, computer hacks and cybernetics', *Media International Australia*, 85(1): 70–8.

Masuda, K. and Haga, S. (2015) 'Effects of cell phone texting on attention, walking, and mental workload: Comparison between the smartphone and the feature phone', *JES Ergonomics*, 51(1): 52–61.

Morikawa, K. (2003) 'Shuto no tanjō: moeru toshi akihabara' [Learning from Akihabara: The birth of a personapolis], Tokyo: Gentōsha.

Morley, D. and Robins, K. (1995) *Spaces of Identity: Global Media, Electronic Landscapes, and Cultural Boundaries*, London: Routledge.

Nigg, C.R., Mateo, D.J., and An, J. (2017) 'Pokémon GO may increase physical activity and decrease sedentary behaviors', *American Journal of Public Health*, 107(1): 37–8.

NISC (National Center of Incident Readiness and Strategy for Cybersecurity) (2016) 'Pokemon toreenaa no minna he onegai' [Points of attention to all Pokemon trainers], 20 July, Available from: https://www.nisc.go.jp/active/kihon/pdf/reminder_20160721.pdf [Accessed 4 January 2021].

Nishidate, A., Tokuda, K., and Mizuno, T. (2016) 'Aruki-sumaho no bōshi ishiki wo takameru keihatsu eizō no naiyō to sono kōka' [Pictures, content, and effects of images that aim to raise awareness for smartphone use while walking], *Research Journal of Teaching and Learning Materials*, 27(0): 109–16.

Nishiwaki, M. and Matsumoto, N. (2018) 'The effects of Pokémon GO playing on daily steps: A retrospective observational study in Japanese male college students', *Japanese Journal of Physical Fitness and Sports Medicine*, 67(3): 237–43.

Obara, T., Kashiwagi, S., and Nakamura, M. (2016) 'Measurement of angles of smart phones at "texting while walking"', Proceedings of the IEICE Engineering Sciences Society/NOLTA Society Conference, 335, Hokkaido University, Japan.

Ono, S., Ono, Y., Michihata, N., Sasabuchi, Y., and Yasunaga, H. (2018) 'Effect of Pokémon GO on incidence of fatal traffic injuries: A population-based quasi-experimental study using the national traffic collisions database in Japan', *Injury Prevention*, 24(6): 485–7.

Ossola, A. (2016) 'You probably don't have smartphone pinky', *Popular Science*, 21 January, Available from: https://www.popsci.com/you-probably-dont-have-smartphone-pinky/ [Accessed 4 January 2021].

Radford, B. (2001) 'The Pokemon panic of 1997', *Skeptical Inquirer*, May/June 2001, Available from: https://web.archive.org/web/20020125093204/http://www.csicop.org/si/2001-05/pokemon.html [Accessed 4 January 2021].

Sanda, D. (2016) 'Pokemon GO launched in Australia and Sydney gamers are everywhere', *The Sydney Morning Herald*, 10 July, Available from: https://www.smh.com.au/business/consumer-affairs/pokemon-go-launched-in-australia-and-sydney-gamers-are-everywhere-20160710-gq2e1v.html [Accessed 4 January 2021].

Sankei Shinbun (2016) 'Geemu ni muchū, mumenkyo bare taiho mo kōtsū ihan tekihatsu 727 ken' [In trance by the game, traffic violations rise to 727 cases – unlicensed driving and arrests], 3 August, Available from: https://www.sankei.com/affairs/news/160803/afr1608030024-n1.html [Accessed 4 January 2021].

Sankei News (2017) 'Poke GO hikinige ni jikkei – kon'yakusha mokuzen de dansei shibō' [Jail sentence for Pokemon Go hit-and-run – man died in front of the eyes of his finance], 27 February, Available from: https://www.sankei.com/affairs/news/170227/afr1702270036-n1.html [Accessed 4 January 2021].

Sanspo (2018) 'Poke GO minagara unten ka jyosei hanerare shibō' [Woman died after getting hit by a driver playing Pokemon GO], 16 April, Available from: https://www.sanspo.com/smp/geino/news/20180415/acc18041522490002-s.html [Accessed 4 January 2021].

Sasao, K., Zhang, J., and Sugamura, N. (2016) 'Development and evaluation of a system using multi-element determination to discourage texting while walking', Proceedings of the IEICE Engineering Sciences Society/NOLTA Society Conference 2016, 227, Hokkaido University, Japan.

Satō, H. (2016a) 'Sekai de mo turaburu ya taiho ga zokushutsu: "Pokemon GO shinya haikai nado de Tōkyō tonai de ha 553 jin hodō"' [Global chain of trouble and arrestations: "Pokemon GO" leading to late-night loitering and 553 cases of arrest in Tokyo alone], *Yahoo! News Japan*, 2 September, Available from: https://news.yahoo.co.jp/byline/satohitoshi/20160902-00061712/ [Accessed 4 January 2021].

Satō, H. (2016b) 'Kyōto de kureensha untenchū ni "Pokemon GO" de jyosei shibō jiko' [Woman died in accident due to crane driver playing 'Pokemon GO'], *Yahoo! News Japan*, 10 December, Available from: https://news.yahoo.co.jp/byline/satohitoshi/20161210-00065347/ [Accessed 4 January 2021].

Sekiguchi, T. and Shun, K. (2016) 'Is self-evaluation for the depth of spatial attention during texting while walking correct?' Proceedings of the 14th Conference of the Japanese Society for Cognitive Psychology, 59, Hiroshima University. Available from: https://doi.org/10.14875/cogpsy.2016.0_59 [Accessed 4 January 2021].

Senoo, A. (2016) 'Pokemon GO no play ritsu ha yaku 4 wari, purei shite mita kansō/okita koto 'batterii no shōkō ga hageshii' "totemo tanoshii" ga jyōi' [Forty percent plays Pokemon GO – among players the foremost impressions are that the game exhausts battery life quickly], *MMD Labo*, 26 July, Available from: https://mmdlabo.jp/investigation/detail_1589.html [Accessed 4 January 2021].

Shūkan Jyosei (2016) '"Nagara pokemon GO" de shibō jiko, 9 sai musuko wo ushinatta chichioya no dōkoku "yo go yo nara, ata wo…"' [Father of the 9-year-old boy who died in an accident where driver played Pokemon GO laments 'if times were different…'], 23 December, Available from: https://www.jprime.jp/articles/-/8773 [Accessed 4 January 2021].

Silverstone, R. and Hirsch, E. (eds) (1992) *Consuming Technologies: Media and Information in Domestic Spaces*, London: Routledge.

Sneep, D. (2019) 'Cell phone city: Reinventing Tokyo urban space for social use', *Asiascape: Digital Asia*, 6(3): 212–36.

Takahashi, T. (2011) 'Japanese youth and mobile media', in M. Thomas (ed) *Deconstructing Digital Natives*, London: Routledge, pp 67–82.

Takahashi, T. (2016) 'Creating the self in the digital age: Young people and mobile social media', in Digital Asia Hub, *The Digital Good Life in Asia's 21st Century*, Hong Kong: Analysis & Policy Observatory, pp 44–50.

Takao, M., Takahashi, S., and Kitamura, M. (2009) 'Addictive personality and problematic mobile phone use', *Cyberpsychology & Behavior: The Impact of the Internet, Multimedia and Virtual Reality on Behavior and Society*, 12(5): 501–7.

Tokushima Shimbun (2016) 'Untenchū ni pokemon GO Tokushima shi futari shishō de yōgisha kyōjutsu' [Tokushima suspect of two casualties due to playing Pokemon GO while driving testifies], 25 August, Available from: https://www.topics.or.jp/articles/-/11185 [Accessed 4 January 2021].

Tokyo Fire Department (2017) 'Aruki sumaho nado ni kakawaru jiko ni chuui!' [Be careful of accidents that can occur due to smartphone use!], Available from: http://www.tfd.metro.tokyo.jp/lfe/topics/201602/mobile.html [Accessed 4 January 2021].

Tokyo Fire Department (2018) 'Kyūkyū hansō deeta kara miru nichijyō seikatsu jiko no jitai' [The status of everyday accidents based on data from emergency medical transport rides], Available from: http://www.tfd.metro.tokyo.jp/lfe/topics/201810/nichijoujiko/data/all.pdf [Accessed 4 January 2021].

Turkle, S. (2005) *The Second Self: Computers and the Human Spirit*, Cambridge, MA: The MIT Press.

Vincent, J., Haddon, L., and Hamill, L. (2005) 'The influence of mobile phone users on the design of 3G products and services', *Journal of the Communications Network*, 4(4): 69–73.

Xie, Y., Szeto, G.P.Y., Dai, J., and Madeleine, P. (2016) 'A comparison of muscle activity in using touchscreen smartphone among young people with and without chronic neck–shoulder pain', *Ergonomics*, 59(1): 61–72.

5

How do Materiality and Corporeality Inform the Intellectual Property Debate? A Case Study of Pirated Media in North Korea

Micky Lee and Weiqi Zhang

Being one of the most politically, economically, and culturally closed countries in the world, North Korea bans all distribution and consumption of foreign media in the country. Therefore, all foreign media are pirated media in North Korea. Despite the state sanction, it has been widely reported that most North Koreans have watched foreign films and television programmes from South Korea and the US (Ahmed, 2014; Yoon, 2015; Hajek, 2017; Baek, 2018). Unlike media piracy in other developing countries, illegal media files are not shared and consumed on the Internet, but smuggled into the country on storage devices such as USB drives. Also, unlike other developing countries, consumers of pirated media can be sentenced to death. The unique situation in North Korea calls for an examination of the material and corporeal aspects of media piracy.

The concept of materiality acknowledges that technologies that create, store, and play media files are not neutral because 'the physical properties or features of objects and settings [...] "invite" actors to use them in particular ways' (Lievrouw, 2014: 23). To give an example,

VCR players do not allow for easy rewinding and fast forwarding, which discourages users from doing so. The technologies with which media copies are produced also give a specific quality to the images. In the analogue era, the quality of pirated media was inferior because quality deteriorates with every dubbing; in the digital era, the quality remains the same even if the original is copied many times.

Corporeality also matters in the case of media piracy in North Korea. Corporeality refers to the physical form of an object, in this case the fleshy, material aspect of the human body. Unlike online media piracy, media piracy in North Korea requires the human bodies to smuggle the physical goods across borders. In this case, media pirates are more like drug smugglers who risk being arrested by bringing forbidden goods across borders. Once the goods are in North Korea, sellers have to hawk the storage devices loaded with media files in a physical space. In addition, since the distribution and consumption of foreign media is a punishable crime, the practice of media piracy brings forth the body that will experience physical pains during punishment. The understanding of corporeality among North Koreans mutually constitutes their understanding of the materiality of media technologies. As we will illustrate later in the chapter, North Koreans who carry a storage device with illegal media have a heightened awareness of the body because possessing such a device constitutes a crime that could lead to corporal punishment.

The mutually constitutive materiality and corporeality of media piracy in North Korea have implications for the debate on intellectual property protection, which tends to see copyrights as legal and economic abstractions. The current debate sees media goods as a commodity so economic reasons, such as profit-making and money-saving, are used to explain illegal file-sharing. This debate also assumes that users from all backgrounds and with different experiences all understand what intellectual property is, why copyrighted material is private property, and why copyright violation is 'stealing' from others. When both materiality and corporeality are taken into account, media piracy is not merely a legal or economic issue, but a set of actions that are technologically embedded and corporeally enabled. Engaging in media piracy in turn gives meanings to technological devices and the body. Therefore, the first part of this chapter delineates how North Koreans understand social relations, space, and their body by participating in media piracy.

However, it has to be noted that information about the everyday lives in North Korea – including about media piracy – needs to be scrutinised because it relies on one single source: North Korean defectors who left the country and now reside in South Korea or the US. In addition, those who are the most willing to talk to the Western press tend to be activists who fight for the reunification of the two Koreas and/or the overthrow of the Kim regime. Therefore, the anecdotes of the distribution and consumption of illegal media in North Korea have to be contextualised in the broader question of why the Western press thinks media piracy in North Korea has news value for its audience.

In the second part of the chapter, we argue that the discourse of media piracy overlaps with that of North Korean defectors. Two overarching but contradictory themes were found in the discourses: First, media piracy is economic and political liberation. Contrary to journalistic accounts of media piracy in other developing countries (such as China), media piracy in North Korea is seen as a brave act committed by North Koreans who want a taste of economic, political, and personal freedom. Second, the starving and dying bodies of North Koreans play a central role in these discourses. Even though the discourses sound convincing, they say North Koreans are starving, yet will allocate their limited resources to buying pirated media rather than food. Why is the starving body so centralised in the discourses, yet is refused importance in the lives of North Koreans?

In the conclusion, we discuss why the discourse of media piracy in North Korea is starkly different from that in other developing countries by situating it in a political-economic context. We argue that the Western press sees North Korea as an antagonistic concept to the West: While North Korea means oppression, tradition, and poverty, the West means freedom, modernity, and affluence. Along these lines, North Koreans are believed to share little universal human experience with Western beings even though they are believed to have the potential to be like Western beings. Foreign media are seen as a catalyst that unshackles North Koreans from the state and assimilates them into Western beings. When North Koreans become Western beings, they are expected to understand the value of private goods. Therefore, even though the discourses of media piracy and defectors highlight the centrality of media technologies and bodies in the acts of media piracy, they also reinforce the superiority of capitalism, in which intellectual property is seen as private property. Therefore, the celebration of North Koreans engaging in media piracy does not contradict the condemnation of media piracy as theft in other developing countries, because Western discourse believes North Korean citizens practice capitalism through the baby step of engaging in media piracy.

Media piracy in North Korea

Media piracy is, on paper, not permissible in North Korea. The country is a member of both the World Intellectual Property Organization (WIPO) and the Berne Convention. Article 44 of the Copyright Act passed in 2001 states that 'institutions, enterprises, organisations or citizens shall not imitate or pirate the works of others that have been submitted for publication'. However, the international community rarely exerts pressure on the North Korean government to stamp out illegal reproduction and distribution of copyrighted materials, first because of North Korea's isolationist stance, and second because foreign media are banned in the country. Therefore, the North Korean government would deny the presence of pirated foreign media in the country.

Similar to some socialist states such as the former USSR and Cuba, North Korea uses the media as a propaganda tool to promote the *Juche* (state self-reliant) ideology and revolutionary sentiment. The government's stance on foreign media reflects collectivism in the country. The state uses the media for nation-building, it condemns media that satisfy individuals' needs and wants. It sees foreign media as a decadent, capitalistic pollutant that harms the collective ideals of the citizens. Consuming or possessing foreign media is a crime against the state and considered treason. When charged, offenders can be sentenced to hard labour camps, prison, or even the death penalty (Freedom House, 2013). However, many bribe police officers to avoid arrest.

Media cannot be illegally shared online, because the Internet is inaccessible to most of the population, except for a small number of officials (estimated to be a few hundred people) (Ko et al, 2009; Baek, 2016). Pirated media mostly comes from China, and loaded on DVDs, USB drives, and SD cards. Some of the storage devices are transported along with legal goods in trucks owned by Korean–Chinese traders who live in China; some are brought into the country by North Koreans, such as government officials who are permitted to travel abroad and are not searched by border patrols (Zhang and Lee, 2019); the rest are sent in balloons or bottles by human rights activists in South Korea, though these tend to be intercepted by the North Korean military (military members may also steal the storage devices and watch the media in secret) (Hajek, 2017).

South Korean TV shows and films are the most popular content, largely because of the common language and shared history. Media content from the US, Russia, and China are also available. According to Greenberg (2015), North Koreans watch mainstream US TV shows such as *Friends* and Hollywood films. Viewers also prefer USB drives and SD cards because they are easy to transport and conceal; users can quickly remove the device and replace it with one loaded with government-approved content during police raids. In particular, a Chinese-manufactured, battery-powered portable DVD player called 'notel' (a neologism formed from notebook and television) is popular among affluent North Koreans because of its low cost and the sporadic electricity supply in North Korea. In addition, SD cards, unlike DVDs, leave no record of what has been watched (Mukherjee and Singh, 2017) so police cannot trace the illegal activities that use them.

Current debate of intellectual property violation in developing countries

The reproduction, distribution, and consumption of illegal and pirated media is certainly not a new phenomenon. In the analogue era, foreign broadcast signals spilled over borders. A good example is of East Germans who lived close to the West German border and could enjoy broadcast signals from the West. Foreign media were also dubbed on audio cassettes or VHS tapes and smuggled across borders. However, the Internet has made distribution more widespread, borderless, and speedy. In particular, BitTorrent sites have been singled out as hotbeds of illegal media sharing activity (Woollacott, 2017).

Even though analogue and digital technologies enable illegal distribution of media files, the current debate on intellectual property violation has not taken into account how specific technologies enact specific practices of media piracy which shape users' understanding of materiality and corporeality. By ignoring the material practices enacted by technologies, the current debate of intellectual property violation tends to narrowly focus on the relations between media piracy, economy, and the laws. The debate also views economy and laws as abstractions in the sense that all economic and legal activities are understood by consumers in the same way in different socio-political environments.

One implication of seeing economy and laws as abstractions is that copyrights are considered private property, a view favoured by media industries and politicians. In this view, intellectual property is more like a house or a car, less like common goods such as air. Therefore, illegal reproduction and distribution of copyrighted content are considered theft. US mainstream media content – in particular Hollywood films – is a popular target. US politicians who tend to protect industry interests openly condemn developing countries for tolerating media piracy among citizens. For example, the US trade representative stated that media piracy is a 'direct threat' to copyright holders. He called out a number of Asian and Latin American countries that tolerate active illicit streaming and sharing of US media content (Robb, 2018). US politicians commonly use confrontational language to discuss these issues, saying developing countries 'abused' Hollywood, claiming the 'thefts' significantly 'cost' the US economy, and vowing strict measures to 'combat' the problem (Fox Business, 2018). Calling the sharing of illegal media theft implies all activities of media piracy can be explained by economic motivations. The same rhetoric is found in South Korea where the Korea Copyright Protection Agency stated that media piracy led to economic loss for the industry (Jang, 2018).

Some scholars disagree with the perspective of US media industries and politicians. However, their counterargument also subscribes to an economic rationale. For example, Karaganis (2011) wrote that even though US transnational corporations framed media piracy as a matter of loss of jobs and profits, law enforcement and education were hardly effective at combatting the problem. Instead, he suggested that the price of legal goods is too steep for consumers in those countries. Therefore, if legal copies were sold at a lower price, consumers would probably prefer them to the illegal ones.

Another counterargument that subscribes to the belief of economic abstraction suggested that politics could interfere in an otherwise self-regulating economy. Studying the case in China, Gao (2014) argued that a non-free market ignores consumers' wants. A state-controlled economy cannot 'contain' taste, so consumers turn to the black market to satisfy their wants (Keane, 2005; Pang, 2006; Li, 2012). In other words, the informal economy is believed to reflect *real* consumers' taste. As we will show in this chapter, the state-controlled economy in North Korea is also said to be a fiction that cannot contain the reality of the black market. Along these lines, Montgomery and Keane (2004) questioned whether a Western model of intellectual property applies to countries where information flow is controlled by governments.

Scholarship on the cultural aspects of media piracy has broadened the media piracy debate by not limiting it to only an economic or legal issue. Castells and Cardoso (2012) suggested that there is little scholarship on how media piracy facilitates the networked cultures of belonging in which social actors use practices and representations to make sense of everyday life. As such, pirates and consumers do not belong to a fringe culture, but create new piracy cultures that point to a new economic paradigm in which consumers develop 'media relationships outside institutionalised sets of rules' (Castells and Cardoso, 2012: 826). As such, media pirates' understanding of an economy and laws is situated in cultural practices, which would include how they understand media technologies and the material culture enabled by them.

Piracy cultures are not underground cultures but constitute a major part of everyday life in politically closed societies. For instance, despite China's economic liberalisation that commenced in the late 1970s, media liberalisation has been uneven (Chan and Qiu, 2002). The limited number of foreign films shown in theatres has led to pervasive media piracy, which then has cultural ramifications on the consumers, such as learning foreign cultures and forming an underground culture of illegal media consumption (Li, 2012; Gao, 2014). The practice of watching pirated media constitutes a private practice in which the audience experiences 'the world through film [...], now an activity belonging to individual time and space' (Pang, 2006: 108). In addition, the audience could form an alternative public sphere in which they discuss films that are banned by the government. Some of them even became filmmakers by learning techniques not taught at official film schools (Li, 2012). As we will explain later, the cultural practice of private viewing is also found in North Korea.

To conclude, the existing literature has not considered how the practice of media piracy is enabled by media technologies that shape an understanding of cultural, economic, and political self. In the following section, we will use media piracy in North Korea to show why materiality and corporeality matter in media piracy.

Materiality matters: The case of media piracy in North Korea

The concept of materiality acknowledges that technologies are neither passive 'things' waiting for human beings to use nor neutral tools that store or transmit media content. Calling attention to materiality allows scholars to look at the interplay and mutual shaping of technological tools, human actions, and social/cultural formations (Lievrouw, 2014). Mukherjee and Singh (2017) have shown how working-class populations in rural India cultivated music sharing on memory cards in the assemblages of phones, networks of young people, and street vendors. Materiality matters in media piracy in both the analogue and digital eras. As shown by Mukherjee and Singh (2017), customers would bring memory chips to 'download vendors' in towns to load pirated files. Then they would insert the chip into a phone shared by the family.

This example shows that the assumption of media piracy being an online activity and file-sharing being immaterial is wrong. Similarly, the case in North Korea shows that the materiality of media piracy relates to social networks that are anchored in local places, and the exchange of technical devices in turn influences how North Koreans see space and the body. We will now explain how this takes place.

First, North Korean consumers buy pirated media from people whom they already know and trust. This differs from online sharing, which assumes the buyers and sellers are anonymous. In North Korea, economic transactions of pirated media are embedded in existing social relations because both buyers and sellers are aware that neither party should report the activity to the police or both will be punished (Kim, 2019). Unlike what economists assume, economic exchange does not always take place between anonymous buyers and sellers (Harrington, 2008). Economic activities are anchored in a local place because North Koreans cannot visit other cities without the government's permission, therefore buyers and sellers probably come from the same geographical area.

Even though smugglers may be motivated by economic interests, buyers and sellers may not see themselves as mere economic actors who enter an exchange to make a profit or acquire goods. Economic transactions may solidify social relations, such as strengthening trust between friends. As such, an economy is not an abstraction in which anonymous buyers and sellers maximise their gains. Seeing pirated media as social goods challenges the dominant view of intellectual property violation, which sees media piracy is primarily an economic activity.

The materiality of media content not only changes social relations between North Koreans, but it also changes how they make meanings of space. Pirates and consumers renegotiate what space means by trading in specific places. For example, Kim (2019) wrote that pirated media are sold in the black market but are not displayed along with legal goods. Buyers have to use a secret code to ask for the goods. As in China (Li, 2012; Gao, 2014), illegal media content is watched exclusively in a private setting or non-cinema businesses such as bars and clubs. However, unlike in China, where electronic goods are affordable for the middle-class, most North Koreans may not find media technologies affordable. Therefore, they tend to consume pirated media in a group in a private space. For example, a defector who was once a North Korean police officer found a group of friends hidden under a blanket in a closet during a raid (Greenberg, 2015). Furthermore, viewers would also deprive themselves of sensory needs by muting the volume when watching the media in a group to avoid attracting attention (Baek, 2016). The act of simply watching but not hearing shapes North Koreans' experiences with foreign media content.

The materiality of media content also changes the meanings of the body because smugglers and consumers have a heightened awareness of the material presence of storage devices. North Koreans have to physically 'feel' the devices and mentally prepare for situations in which the devices need to be concealed. This is unlike online sharing, in which participants are more worried about revealing IP addresses than concealing a technical device. News articles vividly describe how North Koreans mentally prepare for such tasks. A North Korean defector hid DVDs 'under my clothes, and some of them in my jackets, inside the pockets in the jackets' (Hajek, 2017). Owners of storage devices said they needed to mentally prepare to swallow the USB drives if caught by the police (Hajek, 2017). Possessing a piece of immaterial good therefore brings forth the centrality of the body among the audience.

North Koreans negotiate the meanings of these material goods in the context of social relations, space, and the body. These material goods are not passive objects waiting to be used. Rather, they enable North Koreans to shape their private selves that are not prescribed by the state. Li (2012) also suggested that watching pirated media helps the audience form an alternative public self that is not imposed by the state. This private self is cultural, economic, as well as political; at the same time, this private self is mutually shaped by its awareness of the state-approved public self. Culturally, North Koreans are exposed to foreign cultures that are banned by the state. Economically, the goods are acquired in an unregulated market. Politically, they engage in illegal activities that could lead to execution. Does an awareness of a private self undermine the public self imposed by the state? Do the private and public selves co-exist with each other? Does this private self subvert state control (Li, 2012)? Choi suggested that 'there is a very delicate but nonetheless powerful relationship between official 'political-collective' and everyday 'apolitical-private' modes, myths and realities' (2015: 5) in North Korea. Implications of this relationship will be discussed in the following section.

Memory of being a North Korean

Material goods help North Korean defectors articulate their private/ public selves in a new country and culture. Their sense of private selves in South Korea is still mediated through the former public selves in the North. In the absence of digital devices that store and distribute personal memories (such as online photo albums, blogs, and video sharing sites), North Koreans defectors brought with them analogue media (such as photo prints, diaries, and artefacts) to help them remember their private and public lives in North Korea.

A Reuters article 'They escaped from North Korea: Personal stories and mementos of defectors' (Yeom, 2018) asked defectors what they brought with them from North Korea. Some brought photos because: 'even if I died trying, [...] at least I would have this picture with me'. Another brought diaries as 'a record of my history in North Korea', that also serve the purpose of helping to understand North Korean minds when the two Koreas unify. Another defector believed the written pages in her diary prove her existence. When she was hiding in the mountains alone in China, she worried that she'd go insane: 'I read them when I want to remember home. I can't return home, and I already have no memories of my hometown. [....] My diaries are a record of my life. They prove I'm alive'.

Artefacts that are personally meaningful to defectors were also brought along during the escape. A military officer kept his uniform and ID because he deemed them to be 'valuable assets' that would permit him to do anything if he returned to North Korea. However, North Korean military uniforms are not allowed in South Korea so he handed it to South Korean intelligence. A physically disabled defector who left with a pair of wooden crutches refused to throw them away because they symbolise the love he received from friends and family in North Korea: 'the wood from my North Korean [crutches] is hard and painful. But I still keep them, so as not to forget those memories'. The crutches also symbolised North Koreans' yearning for freedom when this defector raised them at President Trump's State of the Union address (Choe, 2018).

The materiality of technical goods – be they analogue media, military uniforms, or physical aids – shapes the North Koreans' private and public selves after they defect. Defectors are not free from a state-designed public self even after they leave the country. For example, the military officer understands that his uniform stands for privilege *only* when he wears it in public in North Korea. The uniform also stands for the authority of the North Korean state so he had to give it up upon entering South Korea. The defector with a physical disability held on to his pair of handmade crutches not only to remind himself of his family in North Korea, but also to publicly demonstrate a single person's political triumph over the totalitarian government.

Interestingly, analogue media were seen to testify to ones' existence better than the body and the mind. One defector believed that if she died during her escape, at least her picture would prove that such a person had existed. Another defector no longer trusted her memory, so written words in her diary reminded her of the past.

In all these cases, the public and private selves are articulated with material goods. Analogue media, military uniform, and homemade crutches help North Koreans articulate their new defector identity through negotiating with their former public selves. Kim (2013) suggested that defectors' narratives are outcomes of their engagement in seeking and conferring meaning in and on their North Korean past and the challenging environment in South Korea. After they left North Korea, their former public selves did not become meaningless, but formed the basis of their new identities in South Korea (Choi, 2015). What is also interesting is that the former public selves are imprinted and carried over in a material form because defectors do not trust their bodies and minds to be reliable forms of memory. We will discuss later in this chapter why bodies play such a prominent role in the defectors' discourses but are yet denied importance in the lives of North Koreans.

News value of media piracy and defectors

We began this chapter by stating that the current debate on intellectual property protection views economy and laws as abstractions, and that it does not take into account how the use of technical devices actively shapes participants' understanding of economic, cultural, and political selves constructed in their social networks and space. As we have shown, the case of media piracy in North Korea indicates that users often did not see themselves as merely economic beings who pay for a commodity or as legal subjects who steal copyright owners' private property (Klinger, 2010).

However, there remain broader questions: From whom do we learn about media piracy in North Korea? Why is media piracy in North Korea considered a topic with news value in Western discourse? What are the readers supposed to learn about North Korea from this discourse? Answering these questions requires an interrogation of the news discourse itself: Who are the news sources that inform everyday lives in North Korea? What are their backgrounds? How do these news sources and Western journalists construct knowledge about everyday lives in North Korea? What are the functions of this discourse? After answering these questions, we will discuss how the ontology of everyday life in North Korea reinforces the dominant view of intellectual property protection.

Methodological challenges to studying North Korea

Researching the everyday life of North Koreans remains an academic challenge. The reliability of official news sources is doubtful because the North Korean government uses the media to glorify the leaders' achievements rather than describing the lives of citizens. Foreign visitors to North Korea are not allowed to interact with citizens and their itineraries are restricted by the government through tour guides. To North Korean researchers who do not read Korean (including the two authors of this chapter), news articles published in the Western press form the basis for knowledge of North Koreans' daily lives. North Korean defectors who left the country with no hope of returning serve as the news sources regarding everyday life in North Korea. Therefore, they do not represent all North Koreans. As suggested, as their sense of selves after defection are shaped by their former public selves, defectors negotiate their old and new selves by talking about their former lives. In addition, those who are willing to talk to the press tend to identify themselves as activists who wish to unify the two Koreas or to overthrow the Kim regime. Some defectors are reluctant to talk to the press because the North Korean government could use 'guilt by association' to punish defectors' family members. Murray (2017) called the over-reliance on defectors as news sources lazy journalistic practice that compromises the reliability of the stories. He even suggested that some defectors, with help from activist groups, fabricated their stories for the Western press. Further, their accounts follow a conventional journalistic form that rarely allows for self-representation (Choi, 2015).

Therefore, we argue that media piracy discourse in North Korea has one single voice. Because of this, *the discourse of media piracy is that of North Korean defectors/activists*, the discourse of media piracy does not exist apart from that of North Korean defectors/activists. It is subjugated to the subject of North Korean defectors/activists in South Korea. This single-voiced discourse in turn constitutes knowledge of everyday life in North Korea. It speaks with authority about a supposedly 'real' or 'hidden' North Korea (Choi, 2015). Lastly, this discourse is mutually shaped by the defectors and the Western press for an English-speaking audience. The press has an agenda about the *topic* North Korea and media piracy serves as a vantage point from which this agenda can be designed and deployed. In the following section, we first examine who the defectors are, then where the media piracy discourse comes from, and finally similar themes between the discourses of media piracy and defectors.

Who are the defectors? How did they leave North Korea?

It is illegal to leave North Korea without the country's permission. Defecting is deemed an act of treason. However, during the Great Famine in the 1990s, the border was more porous, and defectors went to China to look for food and returned to North Korea, crossing the North Korea–China border on their own after paying US$10 bribes to border patrol soldiers. When the economy improved, Kim Jong-un tightened up the border, so defectors have to pay brokers (mostly North Korean defectors) US$10,000 in advance to be transported to Mongolia or South Asian countries (such as Thailand and Laos) where they contact the South Korean Embassy, which flies them to South Korea. There they receive education, job training, subsidies, and housing from the South Korean government under the North Korean Defectors Protection and Settlement Act of 1997. However, it is illegal for defectors to return to North Korea. Once they leave the country, they cannot return. Educated defectors who once had careers in the North Korean regime could work for activist groups or think tanks in South Korea. The unskilled defectors compete with unskilled South Koreans, but tend to face job discrimination because they are perceived as less knowledgeable and culturally backward (Epstein and Green, 2013; Kim, 2013). Suicide rates among defectors are higher than among South Koreans in the new country (Evans, 2015). The less than ideal living situation in South Korea may shape how defectors remember their former lives in North Korea.

Since 2001, the majority of defectors have been women in their 20s and 30s (Kim, 2013). However, based on the news articles collected for this analysis, men are more likely to be interviewed than women. Most defectors came from North Hamgyŏng and Ryanggang Provinces in northeastern North Korea, which border the Yanbian Korean Autonomous Prefecture in China where ethnic Korean-Chinese reside. It is unknown whether easier access to foreign broadcast and cellphone signals lead them to defect. Some defectors stated that outside information made them question the official information disseminated by the government (Baek, 2016). Moreover, residents who live close to the border have more and easier access to foreign media smuggled from China. It can be assumed that defectors are already familiar with foreign media such as Hollywood films and South Korean television shows before defecting.

Where did the discourse come from?

In order to find out how North Korean defectors and Western journalists construct the knowledge about the everyday lives in North Korea and the functions of this discourse, we first point out that the discourse of media piracy *is* that of defectors. In both discourses, some major themes are media piracy, hunger, and death. Even though these themes are seemingly disparate, the discourses connect them by setting up a dichotomy of body and mind.

We gathered the news stories by searching for articles written in English on Google Search with the key phrases 'North Korea and pirated media', 'North Korea and piracy', 'North Korea and black market', and 'North Korean defectors'. We used Google Search instead of an academic database because general readers who are interested in these issues are unlikely to use a specialised database. We collected articles that appeared on the first five pages of search results because most readers are unlikely to go beyond the first few pages for information. After eliminating news briefs with little context, repetitive articles, and irrelevant topics (such as sea pirates), forty-four articles were gathered for analysis, of which sixteen were primarily about media piracy and twenty-eight were about North Korean defectors. Articles came from publications based in the US (*The East Bay Times, Forbes, The Mercury News, Newsweek, The New York Times, USA Today, US News and World Report, The Washington Post,* and *Wired*), the UK (*The Daily Express* and *The Telegraph*), Saudi Arabia (*The National*), and Australia (*The Sydney Morning Herald*). Some articles came from online publications (*MSN.com, The Outline, Quartz,* and *Think Progress*), television news (*BBC News, CNBC, CNN,* and *NBC News*) and public radio stations (KUAF 91.3, KUOW, and NPR). Others came from news agencies (Reuters), government-funded information sites (Voice of America), university student newspapers (*Indiana Daily Student*), and think tanks (The Independent Institute). A small number target a specific audience, such as the military (*C4ISRNET – Media for the Intelligence Age Military*) or Microsoft users (Neowin.com). Despite the origins and audiences of the publications, the discourse does not have diverse views about North Korea. The analysis revealed that all news stories have a similar stance towards issues of media piracy, hunger, and death in North Korea. Two themes stand out in the discourses and will be discussed in the following sections. The first theme is that media piracy is political and economic liberation; the second theme is the starving and tortured bodies of North Koreans.

Theme 1: Media piracy is political and economic liberation

The first theme that is found in both the discourses of media piracy and defectors is that pirated media bring political and economic liberation to North Koreans. In this case, pirated media are not seen as *stolen goods*, but primarily as *foreign* media that enlighten a population that is believed to be duped by the North Korean government. The press believes that enlightened North Koreans will be free from state oppression. However, political, economic, and cultural freedoms are conflated into one: Not only is political freedom conflated with economic freedom, but political freedom and economic freedom are also narrowly defined as personal liberty and capitalistic economic exchange. Further, political liberty and capitalistic economic exchange are believed to bring Western culture to North Korea.

The Western press contrasts North Korean state-controlled media with foreign media by seeing the former as mind-controlling and the latter as liberating. It sees Hollywood films as cultural materials to teach fundamental human values such as love to North Koreans. For example, Park Yeon-mi, a North Korean defector and high-profile activist who regularly appears in the Western press, said *Titanic* taught her that it is possible to 'love' someone who is not the Dear Leader (Greenberg, 2015). She stated that she grew up believing love should be reserved for the leaders and she has never heard her father expressing love for his family. The Western press suggested that North Koreans are not like Western beings until they have a taste of love, which is assumed to be a universal human value. Klinger (2010) also found that the Western press used a positive tone to describe the Afghan audience who watched pirated copies of *Titanic*. The film was believed to bring freedom to the oppressed; and Hollywood was believed to bring news possibilities of social identity and cultural styles to the people.

In the North Korean case, the Western press also suggested that participating in buying and selling pirated media would free North Koreans' minds because a capitalistic market is assumed to be the most natural for human activities. During the Great Famine in the 1990s, the government stopped providing food and other daily necessities to the population. Those who subscribed to a socialist ideal were said to have starved to death after a fruitless wait for ration resumption: 'the intellectuals, bureaucrats and believers in the government died in droves' (*The Bismarck Tribune*, 2018). In contrast, those who did not subscribe to the ideal managed to survive by exercising their economic agency: 'Industrious housewives hawked worn kitchen bowls for handfuls of rice in the black markets. The goal was simple. Secure food for the day, period' (Wee, 2014). The press uses North Koreans' participation in the black market to show that humans are *naturally* entrepreneurial, but this hidden talent needs to be unleashed under a desperate circumstance.

The press also said that foreign media teach North Koreans to be capitalists. For example, news articles suggested that North Koreans hacked state-supplied radios to receive broadcast signals from China, from which they learn to trade (Lee et al, 2008). They learnt about commodity prices and economic conditions from foreign radio. Chinese traders in North Korea were another source for North Koreans to learn about market demands (Baek, 2016; Jeppesen, 2018). The press said that the black market was so successful that it has supplanted the state-controlled market as the *real* market where prices reflected the *real* demand and supply of goods. For example, a South Korean think tank report (Lee et al, 2008) stated that black market prices were already stabilised and traders were making money in the black market prior to the state promoted business practices after the Great Famine. Not only was a capitalistic market system said to keep North Koreans alive (Cho, 2007), but it was also believed to reflect the *real* wants and needs of the average North Koreans. In other words, the *real* North Korean economy is the black market economy, the *fictional* one is the one run by the state.

Economic freedom was believed to bring cultural changes as well. The new generation that grew up during the Great Famine was said to believe in reality rather than ideals, it seeks more autonomy and openness than the previous generations who benefitted from government rations (Cho, 2007). The younger generation also learn about the 'good life' from South Korean television shows with beautiful scenery, nice houses, and cars. These shows train the younger generation to be more consumerist, such as imitating the fashion styles in the shows (Yoon, 2015). They also teach this generation 'Western' practices such as showing affection towards the opposite sex in public (Cho, 2007).

The news discourse set up clear cut dichotomies between capitalism and socialism, individualism and collectivism, freedom and oppression, new and old, wealth and poverty, and reality and fiction (Choi, 2015). Hollywood films and South Korean media – which, despite their highly unrealistic worldviews, are surprisingly said by North Koreans to be about reality – are believed to be catalysts that shift North Koreans from having undesirable traits to desirable ones. These dichotomies are also reflected in publications from the Korean Institute of National Unification, a South Korean think tank. For instance, a publication (Lee et al, 2008) stated that the Great Famine taught North Koreans to have more individualistic perceptions because they were too busy making a living than engaging in political struggle. The lack of contesting views about political and economic freedom in North Korea has implications on foreign policy from other countries. For example, South Korean government set up loudspeakers near the North Korean border to blast loud K-Pop music hoping to persuade border guides to lose confidence in their country (Bacon, 2018).

Theme 2: Starving and tortured North Korean bodies

We have discussed earlier that corporeality matters in media piracy in North Korea because smuggled media in storage devices are physically transported into the country. In addition, North Koreans have a heightened awareness of those devices because they need to be hidden during police raids. The news discourses overwhelmingly described the North Korean bodies as starving and tortured. At one level, the bodies are emphasised because the government is believed to own and control the bodies of its citizens. At another level, the centrality of the North Korean body in the news discourse reinforces the dichotomy between animals and humans ingrained in Enlightenment thought. North Koreans, unlike enlightened Western beings, are said to be more like animals than humans: Their goals are to find food and avoid death. To reach a higher purpose in life, foreign media elevate their minds, which results in North Koreans shifting attention from their starvation to higher ideals.

Academic literature (Ryang, 2000; Kim, 2014) states that North Korean leaders call themselves the parents of the nation who are ready to sacrifice for the citizens. The leaders also called upon loyal citizens to readily sacrifice themselves for the country. The North Korean brand of socialism highlights collectivism in a family and the ethnic group (Kim, 2011). When the citizens are owned by the state, their lives and deaths are not private matters. For example, the outspoken defector Yeon-mi Park wrote that the state has indoctrinated citizens to believe that their lives belong to the state, so dying for the country is the utmost sacrifice that ones can make for the leaders. The ideology was likened to be slavery and abduction: 'we were told in school, in our books, and on television that dying for the Kims was the most honourable sacrifice one could make. [….] North Koreans, in effect, are abducted at birth – and made slaves to the regime' (Park, 2017). The labour performed and embodied by the citizens was also said to belong to the state: '[prisoners] are told [they] had been offered redemption through labor for their country. Instead, they chose to reject the generosity of the country's government. After this they are simply executed' (Ferson, 2012). Suicide is also said to be a means for the citizens to reclaim their bodies from the government. Some defectors, if caught, would choose death rather than having to go back to North Korea. They are said to have carried poison and weapons during their escapes, ready to kill themselves if caught: '[We] armed with knives and prepared to kill ourselves' (Park, 2017); '[a defector] would kill herself rather than be sent back' (Perlez and Lee, 2018). Killing oneself is then a subversive act that North Koreans could perform to rebel against the state.

The hungry country

Hunger and starvation play an essential role in the discourses of media piracy and defectors. First, the Great Famine drove North Koreans to participate in the black market to trade food, illegal media, and other goods. Second, hunger drove North Koreans – some who saw a 'better' life in foreign media – to defect. In both cases, the government is believed to have withdrawn from its parental responsibility because of being a failed state. News articles wrote that the government was unable to 'feed the population' after the famine (Hastings, 2017) even though the current leader Kim Jong-un was quoted to have said that he will 'never […] let the people starve' (Grover, 2017).

Food shortages among average citizens have become the identity of the country to the outside world. It was said to be a 'hungry nation' (Wee, 2014), a 'starving country' (Greenberg, 2015) in which citizens could hardly fully develop physically and mentally: '[hunger stunts] growth in 28 percent of the population' and 'a quarter of North Korean military conscripts are disqualified for cognitive disabilities' (Greenberg, 2015). Hungry citizens were so weak that they were unable to perform daily tasks: '[a defector's] family became so weak that they spent most of the day lying on the floor, sometimes hallucinating' (Choe, 2018).

In the news discourse, starvation made North Koreans equivalent to, or even lower than, animals. North Koreans became desperate because of hunger: '[a defector and his family] fought rats for the seeds stashed in their burrows' (Fishman, 2018). A defector said that in China, 'common animals were eating better than the humans in North Korea' (Fishman, 2018) and 'even dogs in China had food' (Wong and Sagolj, 2018). The Great Famine was said to create such dire conditions that 'in some areas people resorted to cannibalism' (Grover, 2017). Like animals, the primary purpose of North Koreans is said to be finding food.

Also like animals, North Koreans try to avoid deaths imposed on them by the government. Watching foreign media is a crime, so offenders could be sent to prison camps or executed when caught. Ahmed (2014) stated that ten government officials were executed for corruption *and* watching South Korean soap operas. Greenberg (2015) reported that the government executed 80 people, including smugglers of illegal media. Executions are highly performative in North Korea, making a show of the state's ownership of the bodies. Often the state was said to force prisoners to witness executions and bury corpses: 'Once [a North Korean defector] and other inmates were ordered to stone the hanging corpses of would-be escapees' (Greenberg, 2015). The regime not only controls the physiological needs of North Koreans such as eating, labouring, and dying, but it also owns the dead bodies by using them as a public humiliation to show the alive the amount of power that the state has over their bodies.

Body versus mind; food versus USB

In the dichotomy of the West versus North Korea, the West is believed to bring political and economic freedom to starving, dying North Koreans via pirated media. North Koreans, who are described to be animal-like, are said to have brainwashed minds that can only be enlightened by films and television shows produced in capitalistic media systems, such as Hollywood and the Korean Wave. However, the discourse also subscribes to an irrational belief that pirated media are so highly valued in North Korea that North Koreans would forgo eating to acquire those goods. As shown, on the one hand, North Koreans were said to be poverty-striven and their number one goal was to find food to survive. On the other hand, they were said to enjoy illegal media that cost a lot of money: 'a USB stick loaded with contraband films sells for more than a month's food budget for most middle class North Korean families' (Greenberg, 2015). Somehow, food and pirated media became two of the few things that North Koreans really want. Foreign media, along with food, medicine, and American dollars, were sent in balloons and bottles by activist groups from South Korea to North Korea (Voice of America, 2018). One young North Korean defector believed that food was as important as foreign media. Once he was in China, 'he didn't have to beg for food and could watch foreign media, like K-Pop' (Isaacman, 2018). In fact, one defector-turned-activist believed that foreign media are more effective at liberating the population than food aid because the former would change their worldview while the latter would be seized by the state (Baek, 2016). In short, the news discourses stated that food is important to North Koreans as humans, yet it is not as important as pirated media to the lives of *enlightened beings*.

The irrationality of seeing food and pirated media as equally important in the discourse becomes apparent when one takes into consideration how domestic and international aids agencies talk about large-scale hunger in other countries. For example, donors are rarely asked to give pirated Hollywood films to food-insecure children or natural disaster survivors. This irrational belief harboured for North Koreans may mean that *the problem to be solved is not the hunger or poverty of North Koreans,* but the concept of 'North Korea' itself. Hunger and poverty among the populations are just symptoms of this larger problem. Ironically, one solution to solve this problem is media piracy, which the Western press and the US government consider a problem in developing countries.

The superiority of capitalism

We began the chapter by arguing that the US-dominated media industry has been seeing media piracy in developing countries as theft that cuts into the profits of copyrights holders. This argument sees the economy and laws as abstractions, and does not take into account how technologies that enable media piracy shape users' understanding of distributing and consuming pirated media. The case of North Korea exemplifies why materiality matters in media piracy: It shapes North Koreans' understanding of private lives through negotiating social relations, space, and the body vis-à-vis the state. However, the discourses about media piracy in North Korea in the Western press should be critically examined because of two reasons: First, the discourse about media piracy *is* that of defectors. The reason why pirated foreign media is such a central topic in defectors' stories should be questioned. Second, media piracy in North Korea is received positively in the Western press, which contradicts its stance towards media piracy in other developing countries.

The analysis shows that this contradiction actually reinforces the view that the economy and laws are abstractions because the news discourse believes media piracy gives North Koreans an opportunity to learn about capitalism when the communist state fails. The discourse suggested that entrepreneurial North Koreans practised the *most natural* way of survival during the Great Famine, which is buying and selling goods in an informal market. The discourse further suggested that the informal market, unlike the state-controlled one, is an efficient market that reflects consumers' needs and wants. Foreign media content in storage devices, being a sought-after good, has a high exchange value in the market. Being rational actors in a black market, consumers could apply the theory of opportunity cost to determine whether they want food or pirated media. The news discourse sees pirated media as commodities in a free (though informal) market, thus reinforcing the view that the economy is an abstraction.

We have also pointed out one irrational belief in the discourses of media piracy and defectors: Starving North Koreans see food and pirated media as equally important. In some cases, pirated media, not food, is said to be what North Koreans need to liberate themselves from the oppressive state. This irrational belief is however rationalised in the news discourse that subscribes to the dichotomy between North Korea and the West.

Being one of the few communist countries in the world, North Korea is an antagonist of the West. Even though North Korea has little political-economic power to challenge the West, its assumed possession of nuclear weapons and its isolationist stance unsettle the West. As an antagonist, North Korea represents poverty, oppression, and collectivism. In the news discourse, North Koreans are poor, their minds and bodies are controlled by the state, and they are seen as children of the state leaders. These qualities starkly differ from the concept of the West that represents wealth, freedom, and individualism (Shim, 2014). These opposite qualities are not novel, but ingrained in Enlightenment thought that shapes modern nation-states in the West. These ideas also rely on an assumption of racial differentiation by classifying some races as superior and others as inferior. North Koreans who live in the negation of the West are not yet enlightened, they are more animal-like than Western beings. By focusing on basic needs such as looking for food and avoiding death, North Koreans do not share the same human experience as Western beings. Enlightenment thought also believes Western culture to be the standard-bearer of universality, therefore assimilation of North Koreans into Western beings will give them a universal human experience. The tool for this universal experience is pirated media.

There are three problems with seeing North Korea as antagonistic of the West. First, in the news discourses, both North Korea and the West are concepts rather than a country or a specific locale. The concept of 'North Korea' does not necessarily stand for the people who live in the country or culture, but a small number of leaders. This concept reduces the complexity of a nation, history, economy, politics, and culture into a singular entity. Choi (2015) went to the extent of calling the international problem of North Korea a work of fiction that is more than the place, people, and phenomenon. 'The West' is not a more concrete concept than 'North Korea' either. While the US and Western European countries conveniently belong to the West, South Korea – which shares a history, Confucian culture, and language with North Korea – is curiously also seen as belonging to the West in the news discourse. Therefore, the Western press does not attempt to examine how Hollywood films and Korean media are received differently in North Korea because they are all considered 'Western' media in contrast to their North Korean counterparts.

Second, the news discourse uses Enlightenment thought to attribute negative qualities to the concept of 'North Korea' when compared to the West. In contrast, South Korea is a 'Western' country. The Western-ness of South Korea may not be rooted in its culture and traditions, but instead stems from the capitalist political-economic system that formed as it aggressively pursued economic liberalisation after World War II. The news discourse has conveniently ignored that the North was economically more prosperous than the South prior to the Korean War and that South Korea had experienced an economic crisis in the 1990s (see Foreword 2). However, the South Korean government's willingness to be submitted to the West, in particular the US and international organisations, made it more acceptable to the West. In contrast, North Korean leaders, being openly hostile to capitalism and global political-economic influence, are blamed for their inability to manage the country. Pervasive starvation has been used as an example of unsound planning in the problem of North Korea (Hastings, 2016).

Although the Western press sympathises with the average North Korean citizen, binary thinking attributes inferior qualities to North Koreans. North Koreans are said to be passive, malnourished, and poor (Shim, 2014; Choi, 2015), they do not have the capacity to think independently until an external stimulus (foreign media, in this case) triggers them to learn the fine qualities exhibited by Western beings in Hollywood films and South Korean media. To illustrate this, a defector-activist was quoted as saying: 'For every USB drive I send across, there are perhaps 100 North Koreans who begin to question why they live this way. Why they've been put in a jar' (Greenberg, 2015). The image of North Koreans' inferiority is once again reinforced by their lack of Western exposure, which is deemed to be a universal experience. In contrast, food, even though being essential to life, is not considered an external stimulus so it is expected to be less effective at liberating the population. Furthermore, food is also essential to animals so it cannot enlighten a population.

The last and the most major problem with seeing North Korea and the West as antagonistic is that not only are the qualities assigned to both hierarchical, but one quality is *naturally* linked to others. What it means is that a poor country is naturally oppressed, a communist country is naturally collectivist. On the other hand, an affluent country is naturally free, capitalist, and individualistic. As suggested, South Korea, despite being collectivist, is presumed to be individualistic because its media are unlike North Korea's.

The assumption of one quality *naturally* relating to others erroneously conflates political, economic, and cultural qualities. In other words, a political system that ensures personal freedom is believed by default to have a free-market economy in which media goods reflect the culture of the populations. In the news discourse, North Koreans who trade and consume foreign media are seen as not only consumers, but also as activists who rebel against the state's imposed culture, and capitalists who understand how the market works. Conflating the political quality with the economic and cultural ones closes off the possibility that some North Koreans may culturally enjoy pirated media but are satisfied with the current political-economic situation. The discourse also does not recognise that some North Koreans may hope for political freedom but reject the cultural values shown in Hollywood films and South Korean media. Differences among North Koreans are ignored because they are presumed to be passive and incapable of having different thoughts until they are liberated.

Conflating political, economic, and cultural qualities also has an implication on the country's future directions should it reunify with South Korea or initiate liberalisation. North Korea has few alternatives to choose from: If it opens up the market for international trade, it will be asked to ensure political freedom for its citizens; if it allows more political freedom for its citizens, then it will be asked to ensure a free market for foreign investment; if it allows for foreign cultures to be circulated, then it will be expected to ensure political liberty and economic freedom.

By focusing on the materiality and corporeality of media piracy in North Korea, this chapter has refused to see North Korea as a concept that is antagonistic to the idea of the West. We suggested that North Koreans do not necessarily think about the economy and laws as abstractions when they participate in distributing and consuming pirated media, instead they make sense of the foreign content in relation to technologies that help them negotiate social relations, space, and the body. In this sense, what motivates them to participate in media piracy is less important than *how* they participate in it. The question of 'how' acknowledges their agency and creativity in making sense of material goods in their own specific political-economic and cultural environments.

References

Ahmed, B. (2014) '"The Interview," Pirated DVDs, and North Korea's access to Hollywood movies', *Think Progress*, 18 December, Available from: https://thinkprogress.org/the-interview-pirated-dvds-and-north-korea-s-access-to-hollywood-movies-5b9fe34a9a09/ [Accessed 4 January 2021].

Bacon, J. (2018) 'South Korea stops blasting K-pop at North Korea across the DMZ ahead of nuclear talks', *USA Today*, 23 April, Available from: https://www.usatoday.com/story/news/world/2018/04/23/south-korea-k-pop-dmz-north-korea-loudspeakers-nuclear-talks/541292002/ [Accessed 4 January 2021].

Baek, J. (2016) *North Korea's Hidden Revolution: How the Information Underground is Transforming a Closed Society*, New Haven, CT: Yale University Press.

Baek, J. (2018) 'When your body belongs to the state', *Index on Censorship*, 47(4): 36–8.

The Bismarck Tribune (2018) 'Book recounts tales of North Korean defectors', 8 April, Available from: https://bismarcktribune.com/entertainment/books-and-literature/book-recounts-tales-of-north-korean-defectors/article_22ff4c8f-223b-5b4a-80ce-f9639ab05c7c.html [Accessed 4 January 2020].

Castells, M. and Cardoso, G. (2012) 'Piracy cultures: Editorial introduction', *International Journal of Communication*, 6: 826–33.

Chan, J.M. and Qiu, J.L. (2002) 'China: Media liberalization under authoritarianism', in M.E. Price, B. Rozumilowicz, and S.G. Verhulst (eds) *Media Reform: Democratizing the Media, Democratizing the State*, London: Routledge, pp 27–46.

Cho, J-A. (2007) *The Changes of Everyday Life in North Korea in the Aftermath of their Economic Difficulties*, Seoul: Korean Institute for National Unification.

Choe, S-h. (2018) 'North Korean defector, honored by Trump, has a remarkable escape story', *New York Times*, 31 January, Available from: https://www.nytimes.com/2018/01/31/world/asia/north-korean-defector-trump.html [Accessed 4 January 2021].

Choi, S. (2015) *Re-imagining North Korea in International Politics*, London: Routledge.

Epstein, S. and Green, C. (2013) 'Now on the way to meet who? South Korean television, North Korea refugees, and the dilemma of representation', *The Asia-Pacific Journal*, 11(2): 1–14.

Evans, S. (2015) 'Korea's hidden problem: Suicidal defectors', *BBC News*, 5 November, Available from: https://www.bbc.com/news/magazine-34710403 [Accessed 4 January 2021].

Ferson, P. (2012) 'North Korean citizens become more globally aware via piracy', *Neowin*, 26 May, Available from: https://www.neowin.net/news/north-korean-citizens-become-more-globally-aware-via-piracy [Accessed 4 January 2021].

Fishman, M. (2018) '3 North Korean defectors visit Delaware to express thanks', *Associated Press*, 7 April, Available from: https://apnews.com/article/c44ff838361546ac90193260718da821 [Accessed 4 January 2021].

Fox Business (2018) 'Trump to combat movie piracy by China, saving the US economy $600B annually', 15 March, Available from: https://www.foxbusiness.com/politics/trump-to-combat-movie-piracy-by-china-saving-the-u-s-economy-600b-annually [Accessed 4 January 2021].

Freedom House (2013) *Freedom of the Press 2013*, Washington DC: Freedom House.

Gao, D. (2014) 'From pirate to kino-eye: A genealogical tale of film re-distribution in China', in M.D. Johnson, K.B. Wagner, T. Yu, and L. Vulpiani (eds) *China's iGeneration: Cinema and Moving Image Culture for the Twenty-First Century*, London: Bloomsbury, pp 125–46.

Greenberg, A. (2015) 'The plot to free North Korea with smuggled episodes of "Friends"', *Wired.com*, 1 March, Available from: https://www.wired.com/2015/03/north-korea/ [Accessed 4 January 2021].

Grover, J.D. (2017) 'North Korea experiments with freer markets', *Forbes*, 1 June, Available from: https://www.forbes.com/sites/realspin/2017/06/01/north-korea-experiments-with-freer-markets/?sh=2d624ed3f4ae [Accessed 4 January 2021].

Hajek, D. (2017) 'Watching foreign movies is illegal in North Korea, but some do it anyway', *NPR*, 5 July, Available from: https://www.npr.org/2017/07/05/534742750/watching-foreign-movies-is-illegal-in-north-korea-but-plenty-do-it-anyway [Accessed 4 January 2021].

Harrington, B. (2008) *Pop Finance: Investment Clubs and the New Investor Populism*, Princeton, NJ: Princeton University Press.

Hastings, J. (2016) *A Most Enterprising Country: North Korea in a Global Economy*, Ithaca, NY: Cornell University Press.

Hastings, J. (2017) 'How black market entrepreneurs thrive in North Korea', *The Sydney Morning Herald*, April 26, Available from: https://www.smh.com.au/opinion/how-black-market-entrepreneurs-thrive-in-north-korea-20170426-gvsjov.html [Accessed 4 January 2021].

Isaacman, E. (2018) 'North Korean defector shares his story', *Indiana Daily Student*, 4 April, Available from: https://www.idsnews.com/article/2018/04/north-korean-defector-shares-his-story [Accessed 10 February 2021].

Jang, H.Y. (2018) 'South Korea's online piracy paradise', *Korea Exposé*, 5 April, Available from: https://www.koreaexpose.com/south-korea-online-piracy-paradise/ [Accessed 4 January 2021].

Jeppesen, T. (2018) 'A consumer class wields new power in North Korea', *Wall Street Journal*, 1 June, Available from: https://www.wsj.com/articles/a-consumer-class-wields-new-power-in-north-korea-1527867489 [Accessed 4 January 2021].

Karaganis, J. (2011) 'Rethinking piracy', in J. Karaganis (ed) *Media Piracy in Emerging Economies*, New York: Social Science Research Center, pp 1–74.

Keane, M. (2005) 'Television drama in China: Remaking the market', *Media International Australia Incorporating Culture and Policy*, 115: 82–93.

Kim, M. (2013) 'North Korean refugees' nostalgia: The border people's narratives', *Asian Politics and Policy*, 5(4): 523–42.

Kim, S. (2014) 'Mothers and maidens: Gendered formation of revolutionary heroes in North Korea', *The Journal of Korean Studies*, 19(2): 247–89.

Kim, S-y. (2011) 'Dressed to kill: Women's fashion and body politics in North Korean visual media (1960s-1970s)', *Positions*, 19(1): 159–91.

Kim, Y. (2019) 'Introduction: Hallyu and North Korea – Soft power of popular culture', in Y. Kim (ed) *South Korea Popular Culture and North Korea*, London: Routledge, pp 1–38.

Klinger, B. (2010) 'Contraband cinema: Piracy "Titanic", and Central Asia', *Cinema Journal*, 49(2): 106–24.

Ko, K., Lee, H., and Jang, S. (2009) 'The Internet dilemma and control policy: Political and economic implications of the Internet in North Korea', *The Korean Journal of Defense Analysis*, 21(3): 279–96.

Lee, K-d., Lim, S-h., Cho, J-a., Lee, G-d., and Lee, Y-h. (2008) *Changes in North Korea as Revealed in the Testimonies of Saetomins*, Seoul: Korea Institute for National Unification.

Li, J. (2012) 'From D-buffs to the D-generation: Piracy, cinema, and an alternative public sphere in urban China', *International Journal of Communication*, 6: 542–63.

Lievrouw, L.A. (2014) 'Materiality and media in communication and technology studies: An unfinished project', in T. Gillespie, P.J. Boczkowski, and K.A. Foot (eds) *Media Technologies: Essays on Communication, Materiality, and Society*, Cambridge, MA: The MIT Press, pp 23–51.

Montgomery, L. and Keane, M.A. (2004) 'Learning to love the market: Copyright, culture and China', in P.N. Thomas and J. Servaes (eds) *Intellectual Property Rights, Communication and the Public Domain in the Asia-Pacific Region*, New Delhi: Sage, pp 1–35.

Mukherjee, R. and Singh, A. (2017) 'Reconfiguring mobile media assemblages: Download cultures and translocal flows of affective platforms', *Asiascape: Digital Asia*, 4(3): 257–84.

Murray, R. (2017) 'Reporting on the impossible: The use of defectors in covering North Korea', *Ethical Space: The International Journal of Communication Ethics*, 14(4): 17–24.

Pang, L. (2006) *Cultural Control and Globalization in Asia: Copyright, Piracy, and Cinema*, London: Routledge.

Park, Y. (2017) 'How the black market helped me, and others, escape North Korea', Independent Institute, 15 August, Available from: https://www.independent.org/news/article.asp?id=9147 [Accessed 4 January 2021].

Perlez, J. and Lee, S-H. (2018) '"We are ready to die": Five North Korean defectors who never made it', *New York Times*, 25 March, Available from: https://www.nytimes.com/2018/03/25/world/asia/north-korea-defectors.html [Accessed 4 January 2021].

Robb, D. (2018) 'US trade rep finds movie piracy remains "rampant" around the world', *Deadline Hollywood*, 27 April, Available from: https://deadline.com/2018/04/united-states-trade-representative-2018-world-piracy-report-1202378544/ [Accessed 4 January 2021].

Ryang, S. (2000) 'Gender in oblivion: Women in the Democratic People's Republic of Korea (North Korea)', *Journal of Asian and African Studies*, 35(3): 323–49.

Shim, D. (2014) *Visual Politics and North Korea: Seeing is Believing*, London: Routledge.

Voice of America (2018) 'Aid groups send North Korea a message, aid in a bottle', 6 April, Available from: https://learningenglish.voanews.com/a/4334469.html [Accessed 4 January 2021].

Wee, H. (2014) 'How millennials are shaking North Korea's regime', *CNBC*, 12 November, Available from: https://www.cnbc.com/2014/11/12/how-millennials-are-shaking-north-koreas-regime.html [Accessed 4 January 2021].

Wong, S-L.and Sagolj, D. (2018) 'The cold frontier, part one: A journey along North Korea's edge', *Reuters*, 12 April, Available from: https://www.reuters.com/article/northkorea-china-border-porous-special-r-idINKBN1HJ27I [Accessed 4 January 2021].

Woollacott, E. (2017) 'Rights owners list world's biggest piracy sites', *Forbes*, 10 October, Available from: https://www.forbes.com/sites/emmawoollacott/2017/10/10/rights-owners-list-worlds-biggest-piracy-sites/#68fbf80b718f [Accessed 4 January 2021].

Yeom, S. (2018) 'Escape from North Korea', *Reuters*, 12 September, Available from: https://widerimage.reuters.com/story/escape-from-north-korea [Accessed 4 January 2021].

Yoon, S. (2015) 'Forbidden audience: Media reception and social change in North Korea', *Global Media and Communication*, 11(2): 167–84.

Zhang, W. and Lee, M. (2019) 'Black markets, red states: Media piracy in China and the Korean Wave in North Korea' in Y. Kim (ed), *South Korean Popular Culture and North Korea*, New York: Routledge, pp 83–95.

6

Hyperreal Peninsula: North Korea's Nuclear Cinema and South Korea's Digital Revolution

Elizabeth Shim

Drawing from the televisual simulacra of North Korea weapons provocations and projections of regime power, this chapter examines the emergence of a video-mediated nuclear North Korea in the new millennium within the broader frame of networked digital technologies that have facilitated South Korean media flows into the country. Military displays and the more recent emergence of the leadership's nuclear diplomacy can be evaluated as simulation, and is interrogated in the explicit context of a cultural moment when the people of the territorialised and retrenched nation-state of twenty-first century North Korea are receptive to South Korean popular culture and neoliberal productions (see also Chapter 5 about South Korean popular culture in North Korea).

   This chapter highlights the opportunities and constraints of global media and information flows for the newly emerging society of 'transnational Korea' being built on capitalist imperatives and shaping hierarchical relations. In contrast to approaches that exclusively situate North Korea's nuclear weapons in the defence policy of the nation, this chapter reframes the North Korea missile 'crisis' in the era of Kim Jong-un as cultural expenditure, and as part of Pyongyang's comprehensive propaganda network. Within this configuration, military displays simulate state power at a historical moment when South Korea televisual media is the driving force behind prohibited North Korea leisure time. Mediated technologies, and their capacity to meticulously steer the social, illustrate the uneasy relationship between work and play, and state sovereignty and global flow.

North Korea has a long history of developing nuclear weapons, but the policy was not in the foreground until after the death of founder Kim Il-sung in 1994, and was not in public circulation as simulated deterrence in the digital video space until 2003 (Shim, 2017a), when new technologies were already weakening the boundaries of the state (Debord, 2002) and ushering in capitalist media. Prior to the new millennium, North Korea did not release missile images following testing events, a fact obfuscated by outside media's use of substitute images, including images of the wrong missiles, for the August 1998 launch of the Taepodong-1 and the May 1993 test of the Rodong-1 missile (Munhwa Broadcasting Corporation, 1993; 1998).

North Korea's decision to publicise the weapons as a nuclear deterrent against the US became increasingly paradoxical in the post-Cold War years leading up to Kim Jong-un's ascent to power. A general climate of détente followed after multiple security assurances that date back to at least 1994, which restarted in September 2005 with the six-party Joint Statement on nuclear non-proliferation, reached despite tensions that began with the inclusion of Pyongyang under an 'Axis of Evil' by President George W. Bush during his 2002 State of the Union speech. North Korea was also engaging with the South during President Roh Moo-hyun's term of office, when it conducted its first nuclear test in 2006. Under President Barack Obama, North Korea received additional security guarantees and food aid pledges in exchange for suspending nuclear operations.

Since Kim Jong-un fully assumed power in 2012, Pyongyang's televised weapons provocations, and sudden turn to nuclear diplomacy in 2018, have precipitated a material response from the international community and the general framing of the North Korean 'crisis' as the defence policy of a nation under siege. Absent from the discourse on North Korea policy is, firstly, the purpose of the provocations in a country where networked digital technologies have introduced South Korean visual culture to the population; and secondly, any discussion of North Korea's visualisation of war as a digital deterrent, and response, that reinforces state power (Baudrillard, 1995) using cinematic techniques (Kim, 1987) endorsed by the regime. North Korea provocations and nuclear negotiations do not align with rational defence policy, because the regime agreed to a series of security assurances that would have made the country safer vis-à-vis the external world. This chapter thus explores whether North Korea is enhancing media representation to influence its domestic population and the world, by replacing state reality with a network-driven simulation that then serves as a vehicle for shaping subsequent political reality.

Erosion and authoritarianism

North Korea's weapons tests are consistent with Jean Baudrillard's (1994) ideas about simulacrum, implosion, and hyperreality, used to describe a postmodern society of electronic media. Hyperreality refers to the current condition of digital postmodernity where simulacra are no longer associated with any real referent. By definition, simulacra are the signs, images and models that circulate in the digital space, detached from any real material objects, designed to eventually replace the objects of first-order realities. Simulacra, in aggregation, come to embody the novel realm of the hyperreal through the reproducible medium of the virtual. Hyperreality is therefore the by-product of evolving technology, and heir to the society of the spectacle that took hold in twentieth-century capitalist and communist blocs (Debord, 2002).

North Korea's deterministically staged weapons, functioning as simulacra, contain elements of the hyperreal because of state secrecy. Secrecy, and lack of transparency, play an equally important role as do images and spectacles in reifying the hyperreal, as the 2003 invasion of Iraq demonstrated, when US investigators were unable to locate weapons of mass destruction described in the national intelligence estimate of 2002 (Borger et al, 2003). On North Korea, US intelligence agencies continue to report a wide gap between high and low estimates of warheads (Riechmann and Pennington, 2017), ranging anywhere from 30 to 60 nuclear weapons. While North Korean weapons may have a production centre (Hecker, 2004), and are capable of being detonated, the 'reality' of North Korea's weapons are not always verifiable. Rather, their unverifiable images have largely replaced 'reality' as the organising principle of international actions and reactions. Within this configuration, the Yongbyon Nuclear Scientific Research Centre or the Sohae Satellite Launching Station function much like props on a movie set, restoring the reality of North Korea's military on networks like CNN by providing injections of the physical world to keep the simulacrum alive. In this framework, even a catastrophic war on the Korean peninsula that wipes out millions of real human lives would be nothing more than the 'hallucinatory resemblance of the real to itself' (Baudrillard, 1994: 23).

While Pyongyang conceals data pertaining to its nuclear arsenal, reproducible state media images coded as 'threats', coupled with exaggerated rhetoric from the regime, have often been sufficient to mobilise an international response. North Korea's verbal and visual representation of weapons of mass destruction – a vastly inaccessible form of power exchanged between nuclear protagonists (Baudrillard, 1994) – demonstrates the leadership's use of available cultural materials, including the nation's anti-imperial legacy, a claimed history of anti-colonial struggle embedded in the biography of the nation's founder Kim Il-sung, as well as the regime's liberationist ideology – the unifying ideal that calls on the North Korean population to jettison bourgeois ideas (Park and Park, 1990) – to build a new identity that redefines its position in the international community following the 1994–8 famine. North Korea's use of ideologically driven media in state propaganda at a time when ideology is being contested due to marketisation illustrates why state-sponsored nuclear deterrence is also cultural expenditure at the service of propaganda and political spectacle.

Propaganda imagery designed to promote the state ideology of self-reliance or *Juche* takes on various forms, but even at the earliest stages, North Korea founder Kim Il-sung demonstrated a capacity for using films as a basis for establishing power. It was his son Kim Jong-il, however, who dramatically transformed the state's approach to film-based propaganda, integrating his passion for cinema into a *political* strategy, which was subsequently implemented through the political departments of the Korean Workers' Party (Levi, 2015). Categorically, state-sanctioned nuclear announcements, and related weapons imagery that emerged in the new millennium, are part of this strategy.

North Korea's film-based nuclear politics has not been publicly disclosed – a move that would explicitly connect Kim Jong-il's expertise in film to weapons simulacra, and expose the dictatorship's intention to supersede reality with image. To highlight the deception of hyperreality would also undermine cinematic strategy, a strategy the regime in a 1996 treatise on North Korean filmmaking has evaluated as 'scientific', and turns art into 'one of the most powerful *weapons* of the revolutionary struggle', when it is 'applied to revolutionary practice', leading to 'great social unity, and the constant birth of a new human being, the communist' (Hŏ, 1996: 8–13). By definition, North Korea's weapons simulacra fulfil these objectives.

Even before Kim Jong-il came to power in 1994 (and remained in charge until his death in 2011), he had aimed to strengthen and legitimise the power of his family and the image of the regime overseas through film. He not only directed and produced important films including *Sea of Blood* (1969) and *The Flower Girl* (1972), but was also regarded as a film theorist who understood cinema as a virtual platform, an extension of state ideology. 'To revolutionise direction means to completely eradicate capitalist elements and the remaining dogmatism from the real of directing and establish a new Juche-inspired system and methods of directing' (Kim, 1987: 4), he wrote in *The Cinema and Directing*.

In the same book, Kim introduced a theory, credited to his father, about film's potential to shape the future of reality. 'Like the leading article of the Party Paper, the cinema should have great appeal and *move ahead of the realities*. Thus, it should play a mobilising role in each stage of the revolutionaries struggle' (Kim, 1987: 3), the book states, evoking concepts aligned with the postmodern theory of simulation. Two decades later, under Kim's leadership, North Korea announced its first nuclear test that would, in retrospect, 'move ahead of the realities' by the corralling and orchestrating of the hegemonic opinion of the regime.

Kim's theories of film neither predate nor come after Baudrillard's arguments about the blurring of the line between reality and image, the incorporation of virtual environments into physical spaces. Rather, they run parallel to the postmodern view. As filmmaker-statesman, Kim merged nuclear power with the power of television, folding one world into another, erasing the difference between the 'true' and the 'false', 'real' and 'imaginary'. This strategy, first fully executed in 2006 with the country's first public nuclear test, confirms Baudrillard's theory of deterrence: a form of flow that 'circulates and is exchanged between nuclear protagonists exactly as is international capital' (Baudrillard, 1994: 33). Kim's test began a period of provocations substantiated by threats exchanged with the US, a non-war waged from a posture of deterrence. At the beginning of this new era in 2006, Pyongyang's Korea Central Television (KCTV), also adopted a new news format, replacing a plain blue screen with vivid background panels (North Korean Economy Watch, 2012). These policies began to be enforced in an era when an expanding network of markets in North Korea was emerging as a vital source of South Korean media (Hassig and Oh, 2015).

Disavowing capitalist hybridity

North Korean representations of Cold War-style militarisation are taking place in a post–Cold War space witnessing the successful expansion of South Korean cultural industries (Jin, 2016), as South Korea's particular brand of neoliberal capitalism has won the consensus of leftist and right-wing politics in Seoul as the driving force for national unification (Park, 2015). As capital has replaced the state as the subject of national unification in the South, hierarchical relations have been building between the two sides, mediated by diasporic communities in China – an important source of media flows into the North – in addition to South Korea-based North Korean defectors and other South Korean and independent NGOs (Shim, 2015c).

North Korean viewing of capitalist media, originating from a nation of co-ethnics, is inextricably linked to the emergence of a transnational and perhaps even hybridised Korea that has formed in the absence of territorial unity, a new form of Korean unification that has escaped public attention because of a fixation with territory (Park, 2015) and that is inseparable from the capitalist imperatives that shape hierarchical relations among nodal Korean communities (Park, 2015: 11). North Korea's relations to South Korea media are also forged when North Koreans permanently leave their country, lending support to theories of unification that are already driven by the exchange of capital, labour, and ideas (Park, 2015: 3).

China is an inevitable source of media flows into North Korea because of close trade ties. Not only are North Korea guest workers deployed to work in Chinese factories (Shim, 2018a), North Korean traders, crossing the border either with or without state permission, come in contact with South Korean media. At a time when popular culture has merged with the circuits and technology of digital networks (Kline et al, 2003), televisual media is transferred across national boundaries, appearing and circulating in North Korean markets. Technology of Chinese origin is also procured along with digital content, a pattern that has been established for over a decade (Kretchun et al, 2017; see also Chapter 5 about the flows of media and technologies from China and South Korea to North Korea).

The cross-border transactions bringing North Koreans in contact with globally networked flows have eroded North Korea's sovereignty with their consistent growth and significance. The state has responded in at least two ways: first, electing the path of erosion in exchange for durability (Castells, 2009) in its own media representation; and second, walling off the citizenry from outside influences through punishment and military provocations. Both responses, which are explored in this chapter, have defined the current era of Kim Jong-un. The first measure allows North Korea to join the circuitry of the virtual, or the hyperreal, while finding effective ways to employ information technology in order to perform an image of absolute sovereignty. The second policy stages a late modern walling technique that is ironically part of a complex of eroding lines and waning sovereignty (Brown, 2017).

The decline in North Korean sovereignty has been accompanied by tensions between the growing market economy and general regime hostility toward capitalism (Hassig and Oh, 2015). Ordinances passed in 2002, and introduced in lecture material titled 'On Correctly Understanding the State Measure That Has Readjusted Overall Prices and Living Expenses', published by the Korean People's Army Publishing House (Hassig and Oh, 2015), instructed citizens to no longer depend on the government for rations ushered in a new era. The markets supply the population with daily necessities and pirated media. The appearance of markets follows the physical disappearance of as many as 3 million people during the famine, when under Kim Jong-il, human beings and their very survival ceased to be the reason of things (Hassig and Oh, 2015).

Parallel to the market and a key driver of the development is the unbounded rise of corruption. Corruption has given rise to the creation of new documents on how to stamp out crime, and engendered lectures on preventing defections (Hassig and Oh, 2015). Corruption of cadres as it relates to market and migration – as well as bribes that facilitate the flow of media – is mixed in with an increase in non-political crimes (Hassig and Oh, 2015) and free-market aspirations. Corruption has become responsible for the popularity of South Korean television shows in the North, where customs and border control personnel confiscate South Korean content then watch the shows themselves (*Daily NK*, 2016). Changes are taking root, as young people access South Korean videos through memory cards containing South Korean content that can be installed on North Korean mobile phones (see also Chapter 5). The spectacles bring together a disjointed post-famine population (Hassig and Oh, 2015) or motivate defections.

By all appearances, North Korean simulacra – the recorded footage of North Korea's numerous weapons provocations, made widely available on digital networks – seem unrelated to South Korean simulations, that is, cultural content, frequently dismissed in hegemonic media as lacking in quality and sophistication, and not widely or openly consumed in the post-industrial West as it is in East and Southeast Asia. But as long as two simulacra are in play, one committed to reviving revolutionary ideology, the other created to meet the demands of the 'free market', the North Korean people are wedged between two illusions of absolute Korean sovereignty. Since they consume both media, it is worth exploring whether images of nuclear power are being filmed to compete with South Korean simulations, the latter's arrival in the North preceding the era of the regime's digital deterrence.

The Kim Jong-un period

Recent events have vividly demonstrated North Korea's nuclear cinema has not only allowed the state to revive itself through negativity and crisis, but it has enabled a network-driven simulation to serve as a vehicle for shaping subsequent political reality. The return on North Korea's nuclear investments has been multiplied in the digital video space because its state-sanctioned footage feeds into the larger stream of global media flows, enabling the regime to leverage the decoding and orchestration rituals of hegemonic media, as well as the international media's ability to affect the conduct of 'real life' US diplomacy and foreign policy. If this is indeed the case, it lends credence to another theory: North Korea is using media as its main power, but not just its own media. Through simulations of weaponry that can be reprocessed as disinformation, or photo-ops of summit diplomacy, the regime can leverage hegemonic media to create an event worth reporting.

The ways in which North Korea wields simulation as power in contemporary times have accelerated, as has the impact of global media flows. In the survey that follows, North Korean media 'events', defined as creating a reaction among hegemonic media outlets, and often triggering an official US or South Korean government response, are analysed in the contemporary period beginning in 2012, so as to track over time the changing video-mediated behaviour of the North Korean regime and Kim Jong-un. Running parallel to military and political simulacra as practice in North Korea – where the reality of an unverifiable nuclear stockpile has been replaced by ideologically correct, hyperreal, visual representation of weapons – are North Korean encounters with South Korean popular culture. Their impact is assessed through defector testimony as compiled between 2012 and 2018 by the South Korean government think tank Korea Institute for National Unification (KINU). Inter-Korea encounters are also evaluated through North Korea's proprietary cultural output in the same time period: Official North Korean reaction to capitalist media, smuggled footage, news reports, and Kim Jong-un's dramatically altered South Korea policy after the 2018 Pyeongchang Winter Olympics.

Military and non-military provocations

In the first year of Kim Jong-un's rule, before he began to engage in nuclear and missile tests, military posturing and securing the boundaries of North Korea was a priority for him. The regime released images of the new leader at the border village of Panmunjom (Channel A, 2016), on 4 March 2012. In an appearance that established his image as a leader of the military who secures the borders of the Democratic People's Republic, Kim, at the time an unfamiliar figure to most North Korean citizens, orders troops to be on the highest alert at 'the biggest hotspot where gunfire could be heard anytime due to the reckless provocations of the enemies' (Hancock, 2012). Through representation, state media suggests the viewers consume newcomer Kim's image as a leader of the North Korean military, and associate him with the secured boundaries of the specified jurisdiction of the Democratic People's Republic, lines being tested by transnational media flows.

North Korea conducted its third nuclear test on 12 February 2013, after conducting four rocket tests in 2012, including a satellite launch. The test was preceded by a North Korean propaganda video showing the city of New York in flames. The simulacrum is set in a dream sequence, composed of references with no referents, in which a North Korean photographer has fallen asleep, and imagines total destruction in America's largest city (McCurry, 2013). The footage preserves the fiction of an 'adversary', who, just a year prior, offered food aid in return for an end to nuclear tests and uranium enrichment. It also appears to serve as a buffer against more sanguine images of America found in popular films (Shim, 2016a).

There were no nuclear tests in 2014, but a total of eight episodes of short-range or 'Scud variant' missiles were launched. More highly publicised were the trials of two US citizens, Jeffrey Fowle and Matthew Miller, who were freed within the year (NBC News, 2014), along with Korean-American captive Kenneth Bae. The trials were televised and the foreign press was invited to listen to the charges against the prisoners. The charges were vague, but the incidents helped to elevate North Korea's profile as a threat to US national security, and to the safety of US citizens, as did the cyberattack on Sony Pictures in November. The hack, retaliation for the film *The Interview* (2014) – itself a projection of a hyperreal Korean Peninsula that blends reality and fiction with a Hollywood Kim Jong-un – manifested as cyberwar that unfolds in an abstract electronic and informational space (Baudrillard, 1995), the same space in which global movie business capital circulates.

The cyber breach also elevated North Korean sharp power in the context of international soft power media flows. The Obama administration concluded that North Korea sponsored the attack – a virtual attack that lead to a real outcome in the form of the cancellation of the film's premiere (BBC News, 2014). North Korea's alleged pre-emptive attack of a cinematic event that purely existed on the level of spectacle and hyperreal Hollywood, the ultimate ambiguousness of the attack's origin, and the overwhelming hegemonic response to the hack all congealed in 2014 to link a strategy of 'clean' war to a regime that builds its power on pictures and footage of weaponry (Baudrillard, 1995: 40) and through cyberattacks that unfold in the virtual space rather than through the use of weapons-grade plutonium against an enemy. The uploading of pirated copies of the film to file-sharing services and the more than 900,000 downloads of *The Interview* by Christmas Day 2014 (Spangler, 2014) filled the movie, a 'run-of-the-mill Seth Rogen comedy', with new significance. The mass proliferation of copies augments the situation external to the film, rather than the film's essence. Piracy and public uproar reified the film as a 'rallying cry for free speech' in the US (Isaac, 2014).

North Korea increased military provocations the following year, firing short-range missiles in February 2015 (Center for Strategic and International Studies, 2019). The country then became the centre of global media attention after a landmine incident at the border maimed two South Korean soldiers. The technological simulacra of 'war' that accompanied the incident – footage of a tense Korean Demilitarised Zone (DMZ) and the recovery of the detonation device later shown at a Seoul defence ministry press conference – became not only an integral part of North Korea's operational procedures, but South Korea's as well. The incident was followed by a declaration by Kim that North Korea would use nuclear weapons if provoked (Shim, 2015b), but tensions stopped when officials from both sides met unexpectedly in Panmunjom.

In 2016, the final year of the Obama administration, North Korea continued to engage in acts previously condemned by the international community, including a fourth nuclear test on 6 January and an unprecedented number of tests of submarine-launched ballistic missile tests (Center for Strategic and International Studies, 2019). A good portion of North Korea's tests in 2016 failed and so were not made public, but an August test of a submarine-launched ballistic missile was made public for domestic consumption. North Korean state media connected the launch at sea with the figure of Kim Jong-un, who was quoted in KCNA as saying that North Korea had become a 'premier military power with perfect nuclear strike capability', a response to the 'nuclear blackmail' of the US (Shim, 2016b). The distinction of what this nuclear programme implied was lost in the hyperreality of North Korea's rhetoric and the grin on Kim Jong-un's face.

The following year, the first year of Donald Trump's presidency, North Korea televised two long-range intercontinental ballistic missile launches in July and a ballistic missile flew over Japan. Another ICBM test in November was given a hyperreal afterlife in a music video and through footage of citizens watching the show on a giant LED screen in Pyongyang. Intercontinental or long-range missiles are more provocative, because they are potentially capable of reaching the US and, perhaps more importantly, create a disturbance that can ripple through US media. But in August, North Korea also released photographs of Kim being briefed in a situation room about Guam. The satellite image of the US territory was outdated (Shim, 2017b), possible evidence of lack of real war preparation. Western media largely missed this detail, because of the decision to almost exclusively frame North Korea as a threat (Berlinger and Lendon, 2017) part of a hegemonic preoccupation with decoding and orchestration rituals that demarcate 'periphery' from 'centre'.

Trump's vow at the UN General Assembly on 19 September 2017 to 'totally destroy' North Korea erased the distinction between actual policy and a more real-than-real war. Escalating rhetoric on both sides created unprecedented levels of fear in South Korea, where the US military was reported to be rehearsing evacuation drills (Haltiwanger, 2017). The rise of dangerous US rhetoric, combined with North Korea's increased use of digital deterrence, reached a new climax. Unlike Obama, Trump – who had a background in the world of hyperreality TV, and was elected not on the merits of his rational arguments but rather because of his challenge to competing discourses, including the media – is too volatile for North Korea's deterrence to circulate and be exchanged between stable nuclear protagonists.

The Trump–Kim tensions, themselves a simulacrum, and therefore misrepresentation, never advanced to the use of force in a world system where the virtual has overtaken the actual and functions to deter the real event of nuclear war (Baudrillard, 1995). By the end of 2017, North Korea did not and could not follow through on a threat to detonate a hydrogen bomb (Haltiwanger, 2017) over the Pacific Ocean. It also registered a change of tone after sending its athletes to the 2018 Pyeongchang Winter Olympics at the invitation of the pro-engagement President of South Korea, Moon Jae-in.

The Olympics and a series of inter-Korea concerts in Seoul and Pyongyang, preludes to Kim's meetings with world leaders, set the tone for the remainder of 2018, marked by the absence of military provocations. The North Korean leader's historic meeting with Chinese President Xi Jinping in March, and with President Moon at Panmunjom on April 27 also built momentum for the Singapore Summit on June 12. The historic US–North Korea summit was highly publicised in North Korean state media and boosted Kim's image as a leader.

Diplomacy replaced provocation as Kim met three times each with Xi and Moon in 2018. Kim's meetings reproduced diplomacy as televisual simulacra while constructing the North Korean leader as a statesman who engages with the world. The media spectacles the summits created introduced new cultural codes in North Korean statecraft. Kim's diplomacy eased tensions and reduced perceptions of North Korea state violence among South Koreans (Shim, 2019). The Trump–Kim joint statement signed in June gave rise to denuclearisation as a hyperreal concept. Kim's meetings with Moon and Xi also produced signs that referred only to themselves, such as 'peace' and 'blood-forged alliance', respectively. When North Korea returned to missile tests in May and July 2019, diplomacy was reified as a stylised, selective misrepresentation, or simulacra. It had no referents in the real for a regime that cannot be at total peace or total war with rivals.

Popular culture

Beyond the spectacle of missile provocations, the era of Kim Jong-un has also launched more appealing initiatives into soft power, a bid to create an image of 'material well-being of the regime' (Lim, 2017: 602) that is in the making as the state struggles with non-ideological capitalist media. In 2012, Kim made a public appearance at a televised gala featuring Disney characters, women musicians of the Moranbong Band in mini-skirts, and Winnie-the-Pooh. The deliberate use of trademarked icons mirrors media flows into the country, where, according to defector testimonies, people have underground access to American films including cartoons (Shim, 2016a) and South Korean materials.

Outside the state, diffuse spectacles of South Korean capitalism were reaching almost every region of the world. Digital technologies and the YouTube platform brought unprecedented attention to the music video 'Gangnam Style', released on 15 July 2012, which drew more than 1.6 billion hits on YouTube by 16 June 2013 (Jin, 2016: 152), owing to the rising influence of social media and online 'crowds' as a means of transmitting popular music. The song's popularity made it South Korea's most successful export. Footage smuggled out of North Korea showed ordinary North Koreans watching the 2012 video, along with South Korean dramas featuring actress Song Hyc-kyo, possibly at grave risk to themselves: Figures in the dark watch 'Gangnam Style' and other videos under bed covers, with the shades down, as a dog outside the house begins to bark to warn of approaching strangers (TV Chosun, 2013).

Neither bribes nor official punishment for watching illicit media are sufficient to put an end to the hyperreality of South Korean simulation. South Korea-based North Korea news site New Focus International reported in May that young female university students in Pyongyang were wearing 'hot pants', a reference to mini shorts, and the trend was related to images of women depicted in 'Korean wave' videos, according to the report's source (*Seoul Shinmun*, 2012). The trend is particularly concentrated in communities where women live in groups, such as dormitories.

In 2014, the third year of Kim Jong-un's rule, the North Korean leader again gave visibility to the Moranbong Band, the all-women's group that debuted in 2012. The group played at a concert in April, putting on what was described in the hegemonic press as a 'Western-style pop-rock show' because they performed the theme song to the movie *Rocky*. The performance was made available globally through digital media. In South Korea, however, the band draws comparisons to the popular group Girls' Generation, which is well known in North Korea (Radio Free Asia, 2014) and whose popularity is reflected in Kim Jong-un's later decision to invite Seohyun, a member of the group, to Pyongyang in 2018 (DongA.com, 2018). A year after Kim Jong-un permitted the Moranbong Band to make a public appearance after a three-year absence, a new 'light music choir and orchestra' debuted in July. The Chongbong Band engaged in cultural music diplomacy on a trip to Moscow.

Outside North Korea, South Korean media continued to attract viewers across Asia, and the year's most successful South Korean television event, 'My Love from the Star', became the most expensive drama to be sold in China, at US$35,000 per episode. The 2014 show made its way across the border to North Korea, where women with relatively powerful husbands asked them to procure recordings (*Daily NK*, 2016). South Korea's hyperreality, constructed in the entertainment industry centred in Seoul and promoted by the South

Korean government, is cited as constructing South Korean male sex appeal among North Korean women, and the media spreads the construct of Southern male sex appeal.

North Korea made its struggle with foreign media known publicly in a July 2015 issue of the *Rodong Sinmun*, referring to media only by the euphemism 'cultural decadence' and 'toxins of capitalism' (Shim, 2015a). It was a rare public admission for North Korea, and was not made so explicitly again in 2015. Reports also suggested that young people were dyeing their hair to emulate South Korean stars and that the South Korean style of speaking was becoming trendy in Pyongyang. The trends exemplified the generation by models of the real world, models without origin or reality – a hyperreal world (Baudrillard, 1994) – because of North Korean inability to access South Korean consumer culture at its geographical point of origin or within the proper context.

The infiltration of South Korean popular culture, most prominently K-pop music that was not permitted to be played in the North, was a worry for the regime in 2018, despite Kim Jong-un's hosting of South Korean pop artists, including the girl band Red Velvet, in Pyongyang. In August the North Korean leader condemned 'reactionary bourgeois culture' and Korean Workers' Party newspaper *Rodong Sinmun* said a 'mosquito net' (Yonhap TV, 2018) must be put in place to combat bourgeois ideology, a reference to a transnational global flow that includes K-pop. The popularity of South Korean group BTS has been confirmed by defectors and government officials in the South, who say North Koreans secretly stream BTS music videos on their phones and use a code word to refer to the group.

Defector testimonies

The North Korean state's concerns regarding the population's increased access to transnational media flows are reflected in the use of the death penalty and other forms of violence, to end access to South Korean popular culture. Data provided in 1,263 defector testimonies on media experiences between 2012 and 2018, as compiled by KINU, shows that violence increased over time in the era of Kim Jong-un.

At the beginning of the third Kim era, media of various national origins circulated in the country, but North Korean authorities were reported to have been the most sensitive to the buying, selling, and viewing of South Korean media. While the viewing of unapproved Chinese media required offenders to pay a bribe-fine of 1,000 yuan (about US$150), the viewing of South Korean films carried twice the penalty (US$300) (KINU, 2013). Inability to pay the fine for viewing South Korean or US programmes, resulted in a six-month sentence at a re-education camp.

Evidence is also growing that in North Korean society, South Korean media – sought out in Asia for their strong emotional impact on viewers (Jin, 2016) – have had an irreversible impact on sociality. In 2014, a defector in the South said a family of four grown accustomed to listening to South Korean radio in Sukchon, North Pyongan Province, had been sent to a political prison camp on 23 July 2013, after the head of the household said that 'South Korea is good' in his workplace. A 2013 survey of defectors suggested several million North Koreans regularly watch South Korean television programmes, with up to 70 per cent of the population gaining some kind of access to various South Korean media products through videos (Hassig and Oh, 2015).

The relatively superior television productions from South Korea in an information-poor and propaganda-rich environment are also circulating at a faster pace, owing to the adoption of new technologies. A defector tells South Korean news channel YTN that flash drives are in demand because they are easy to hide, thereby decreasing the chances of arrest (YTN, 2013). Then in a statement that belies the threat transnational flow poses to the regime the defector adds that they do not know why the media is being banned when it 'contains no ideology'. Another defector, Oh Kum-sun, confirms that men and women, young and old alike are 'deeply captivated' by South Korean television (VOA Korea, 2013), and the cost of purchasing the programmes are shared among groups who watch the films together. Memory cards containing South Korean content that can be installed on North Korean mobile phones are also cited as a reason for transnational flow acceleration.

The year 2013 also witnessed the rise of pardons, especially when the offenders were a relatively large group of people, according to KINU data. In Sinuiju, North Pyongan Province, a public trial involving 300 people ended with a pardon, after the group shouted, 'Long Live Kim Jong-un!' (KINU, 2014) A similar reprieve took place the same October at an airfield in Hyesan, Ryanggang Province, where two men were to be executed for providing a space to watch South Korean videos. Both defendants performed regime loyalty, to which the state responded with a show of official clemency. The executions, a simulated performance of absolute state power, restored the reality that had escaped the state amid eroding authority under marketisation.

As a virtual war on a Hollywood movie studio played out in the world outside North Korea's borders in 2014, the regime waged a real war against transnational media at home, a conflict that had real victims, in the form of people sentenced to prison and other punishments. A 2014 KINU survey of 221 defectors showed the secret viewing of South Korean media was 'spreading' and 'cheap recording devices from China' were being smuggled into the country, giving rise to sentiments critical of the regime (KINU, 2014). Bribery continued to be a means of avoiding labour camp sentences, but punishment was meted out to offenders who could not pay their way out.

There is little evidence to believe that transnational media flows have stopped, and a 2015 survey of 186 defectors showed a more serious crackdown had been taking place in recent years, when authorities in each province began to execute two people per month to warn against media viewing and 'trafficking' (KINU, 2015). The 2015 survey also revealed that mobile phones, believed previously to have only used for domestic communication, were being used to 'connect to the outside world in a limited way' for North Koreans able to use Chinese phones that can pick up signals along the border. The spectacle of South Korean wealth translates into defections, posing a grave problem for Kim. A separate survey that includes US data from 2015 showed 34 per cent of defectors said they could access Chinese and South Korean broadcasts through their televisions in North Korea (Kretchun et al, 2017). The state's crackdown against media access continued, and the labour sentence has now increased from six months to a year. Bribes now work after imprisonment.

Thae Yong-ho, the senior Pyongyang diplomat who defected from London, told South Korean journalists in 2017 that soft power should be deployed to counter North Korea hard power, playing on a theme of previous interviews (Chosun.com, 2017). His view underscores the parallels between two hyperrealities, although he does not characterise the weapons in question as belonging to the order of the hyperreal. A US-based defector said in October that the North Korean leadership knows the South can always 'swallow' the North because of its superior economic power, but he suggested the threat South Korea poses is primarily of a military nature (Shim, 2017c).

In a 2018 interview, a defector and radio personality host said a regionally popular South Korean TV drama, *Stairway to Heaven* (2003), had a significant impact on North Koreans in the northern province of Ryanggang, and said it would 'not be an exaggeration' to say nearly all defectors were motivated to leave their country after viewing the soap opera. She also said she watched South Korean drama 'Heirs' and listened to the music of Girls' Generation in North Korea (*Seoul Shinmun*, 2018).

Discussion

The analysis of contemporaneous media events shows North Korea's soft power initiatives did not overturn established transnational media flows originating from South Korea. Kim Jong-un's launch of his own soft power initiatives, including all women's bands that mirrored South Korean pop culture trends, did not change the patterns of prohibited North Korea leisure time, even as North Korean popular culture, with its imitative properties, was championed publicly at an event in state media. In the face of complications and the increased danger of losing an unprecedented amount of control, Kim instead engaged in highly public military provocations with increasing frequency from 2012 to 2017 to create tensions and further isolate the population, a move that could be used to justify his authoritarian rule. In 2018, the regime exchanged its policy of the militarised hyperreal for Kim-centred summit diplomacy in the face of a new US president who subsumed the spectre of nuclear war into his own landscape.

The very public nature of these events in an era when people cease to correspond perfectly to national borders that are being tested by transnational media flows is in some ways a sign of a state in panic, declaring permanent emergency powers to suspend the natural law of flows, and 'face down enemies everywhere', (Brown, 2017: 770) while commanding the execution of non-political dissidents who have become politicised for their association with non-ideological media. For a dictator whose power over his own people takes precedence over everything else (Baudrillard, 1995), there is little he can do, except to continue to produce and restore the real (Baudrillard, 1995: 23), and revive rule through negativity and crisis, even as he must appear, at times, a statesman conducting diplomacy.

The frequently expressed bewilderment of defectors regarding the ban on South Korean media elevates the discourse. South Korean television dramas are described as containing 'no ideology' (VOA Korea, 2013), but North Korea's state-founding ideology effectively monopolises the nation's policies. The ideological vacuum that is South Korean entertainment poses a threat to North Korea, because North Korea also operates on spectacle, owing to its communist past. North Korea's spectacle is concentrated, and much like its early peers the USSR and China, it relied on coercive bureaucracies to accumulate images to be deployed in the cult of personality (Debord, 2002).

An image or story that denies the need for the cult or the dictatorship of the bureaucratic economy, then, be it ideological, non-ideological, anti-ideological, or post-ideological, destroys the bureaucracy completely because it represents an external choice, an option that can potentially channel a vision of transnational unification under contemporary South Korean capitalism. Even an ideological vacuum can be a sign to the citizens that the bureaucracy does not hold a monopoly over wholesale survival, as the central leadership claims.

This survey shows punishment and state-sanctioned violence against viewers and distributors of illicit media increased over time, particularly as North Korean media representations of weapons tests surged by 2017. Similarly, in 2018, punishment for media access and viewing did not reduce (Shim, 2018b) even as Kim appeared outwardly to be more open to outside media and South Korean popular culture. Rather, after North Korea's nuclear cinema was suspended, diplomacy in 2018 continued the pattern of state revival through the use of reliable hegemonic media to spread disinformation – simply another dimension of a consistent strategic embrace of the logic of simulation (Baudrillard, 1995).

This survey also illustrates the irony of North Korea's late-modern walling techniques. The regime's public condemnation of its US captives from 2012 to 2017 for 'anti-state' acts took place when the boundaries between North Koreans and the rest of the world were becoming part of a complex of eroding lines, made possible by increased marketisation and transnational media flows, and subversive acts that were classifiably anti-state were taking place daily within the national interior.

Powers unleashed by globalisation, South Korea's leveraging of globalisation, and its distinctive harnessing of neoliberal capital, could be posing the gravest threat to North Korean sovereignty, and largely because South Korea is a neighbouring nation of co-ethnics who offer a globally networked sovereignty of homeostasis.

Lastly, the findings show North Korea's weapons distinctly serve as a digital hyperreal deterrent, because their real status is still unknown. The international community, the US and South Korea in particular, has spent substantial amounts of political capital in response to a simulation that has become inseparable, unsurpassable from so-called reality, as North Korea's digital deterrent has mostly yielded only one kind of response from US-led hegemony – one that envelops North Korea in signs that denote 'threat'.

North Korea's manoeuvres in the televisual space, designed to elicit such responses, reveals how the regime 'systematises the ideology and theory of the revolutionary tradition of cinematic arts' (Hŏ, 1996: 8) in a counter-hegemonic process that coincidentally disrupts the neoliberal capitalist present and the geopolitics of the Korean Peninsula. In our current state, North Korea's simulacra remind us we are in a place, to quote Baudrillard, where 'we will no longer even pass through to the "other side of the mirror" […] the golden age of transcendence' (Baudrillard, 1994: 125), and where the very probability of war, 'its credibility or degree of reality, has not been raised even for a moment' (Baudrillard, 1995: 67). In the presence of such uncertainty, North Korea's hyperreal claims of nuclear power guarantee that what is effective is the simulation of nuclear power, never the reality.

Acknowledgement

This chapter was made possible in part by a 2019 research travel grant from the Korean Collections Consortium of North America. Special thanks to Hyoungbae Lee and Andrew Johnson of the East Asian Library at Princeton University for their assistance in finding key publications.

References

Baudrillard, J. (1994) *Simulacra and simulation*, Ann Arbor: University of Michigan Press.

Baudrillard, J. (1995) *The Gulf War Did Not Take Place*, Bloomington: Indiana University Press.

BBC News (2014) 'New York premiere of Sony film *The Interview* cancelled', 17 December, Available from: https://www.bbc.com/news/entertainment-arts-30507306#:~:text=The%20New%20York%20premiere%20of,said%20it%20had%20been%20shelved. [Accessed 5 January 2021].

Berlinger, J. and Lendon, B. (2017) 'Next target Guam, North Korea says', *CNN*, 30 August, Available from: https://edition.cnn.com/2017/08/29/asia/north-korea-missile-launch-guam-threat/index.html [Accessed 5 January 2021].

Borger, J., MacAskill, E., and Winter, P. (2003) 'The hunt for weapons of mass destruction yields – nothing', *The Guardian*, 25 September, Available from: https://www.theguardian.com/world/2003/sep/25/iraq.iraq [Accessed 5 January 2021].

Brown, W. (2017) *Walled States, Waning Sovereignty*, Cambridge, MA: The MIT Press.

Castells, M. (2009) *The Power of Identity* (2nd edn), New York: Wiley-Blackwell.

Center for Strategic and International Studies (2019) 'North Korean missile launches and nuclear tests: 1984-present', Available from: https://missilethreat.csis.org/north-korea-missile-launches-1984-present/ [Accessed 5 January 2021].

Channel A (2016) 'Kim Jong Un, 2012nyon Panmunjom pangmun sashil konggae' [North Korea's Kim Jong-un Reveals Visit to Panmunjom in 2012], 21 January, Available from: https://bit.ly/2CiZnLz [Accessed 5 January 2021].

Chosun.com (2017) ' "Namhan norae tŭtko shipta" …moksumkkaji kŏlge han "sop'ŭt'ŭp'awŏ"' ['I want to listen to South Korean songs' … risking their lives for 'soft power'], 21 November, Available from: http://news.chosun.com/site/data/html_dir/2017/11/21/2017112103274.html [Accessed 5 January 2021].

Daily NK (2016) 'Pukganbu anae, "pyŏresŏ on kŭdae kuhaedallat" ttessŭndanŭnde' [Wives of North Korean officers nagging their husbands for a copy of 'My Love from the Star'], 15 November, Available from:: http://www.penews.co.kr/news/articleView.html?idxno=2916 [Accessed 5 January 2021].

Debord, G. (2002) *Society of the Spectacle* (K. Knabb, trans), London: Rebel Press.

DongA.com (2018) 'T'albuk kyosu sonyŏshidae pukaesŏ in'gi t'op… sŏhyŏn (ch'uryŏn), puk ch'ŏngt'agil kŏtt'e' [North Korean defector professor: 'Girls' Generation most popular in North Korea … Seohyun's [appearance], most likely North Korea's request'], 12 February, Available from: http://www.donga.com/news/article/all/20180212/88627205/2 [Accessed 5 January 2021].

Haltiwanger, J. (2017) 'North Korea threatens US with "strongest hydrogen bomb test Over the Pacific Ocean"', *Newsweek*, 25 October, Available from: https://www.newsweek.com/north-korea-threatens-us-strongest-hydrogen-bomb-test-over-pacific-ocean-692848 [Accessed 5 January 2021].

Hancock, P. (2012) 'North Korea's Kim Jong Un visits DMZ for the first time', *CNN*, 5 March, Available from: https://edition.cnn.com/2012/03/05/world/asia/north-korea-dmz/index.html [Accessed 5 January 2021].

Hassig, R. and Oh, K. (2015) *The Hidden People of North Korea* (2nd edn), Lanham, MD: Rowman & Littlefield.

Hecker, S. (2004) 'Hearing on "Visit to the Yongbyon Nuclear Scientific Research Center in North Korea"', Washington, DC: Senate Committee on Foreign Relations, Available from https://fas.org/irp/congress/2004_hr/012104hecker.pdf [Accessed 5 January 2021].

Hŏ, Ŭ-m. (1996) *Yŏnghwa yesul ŭi hyŏngmyŏng chŏnt'ong* [The Revolutionary Tradition of Film Art], Pyongyang: Munhak Yesul Chonghap Ch'ulp'ansa.

Isaac, S. (2014) '"The Interview"', *Plugged In*, 25 December, Available from: https://www.pluggedin.com/movie-reviews/interview-2014/ [Accessed 5 January 2021].

Jin, D-Y. (2016) *New Korean Wave: Transnational Cultural Power in the Age of Social Media*, Champaign: University of Illinois Press.

Kim, J. (1987) *The Cinema and Directing*, Pyongyang: Foreign Languages Publishing House, Available from:: https://www.korea-dpr.com/lib/209.pdf [Accessed 5 January 2021].

KINU (Korea Institute for National Unification) (2013) 'White paper on human rights in North Korea', Korea Institute of National Unification, Available from: https://bit.ly/3cYkPcq [Accessed 9 February 2021].

KINU (Korea Institute for National Unification) (2014) 'White paper on human rights in North Korea', Korea Institute of National Unification, Available from: https://bit.ly/3a70JuJ [Accessed 9 February 2021].

KINU (Korea Institute for National Unification) (2015) 'White paper on human rights in North Korea', Korea Institute of National Unification, Available from: https://bit.ly/2OeDQwI [Accessed 9 February 2021].

Kline, S., Dyer-Witheford, N., and De Peuter, G. (2003) *Digital Play: The Interaction of Technology, Culture, and Marketing*, Montreal, QC: McGill-Queen's University Press.

Kretchun, N., Lee, C., and Tuohy, S. (2017) *Compromising Connectivity: Information Dynamics Between the State and Society in a Digitizing North Korea*, Washington, DC: Intermedia.

Levi, N. (2015) 'Kim Jong Il: A film director who ran a country', *Journal of Modern Science*, pp 155–66.

Lim, T.W. (2017) 'State-endorsed popular culture: A case study of the North Korean girl band Moranbong', *Asia and the Pacific Policy Studies*, 4(3): 602–12.

McCurry, J. (2013) 'North Korea propaganda film depicts New York in flames', *The Guardian*, 5 February, Available from: https://www.theguardian.com/world/video/2013/feb/06/north-korea-new-york-flames-video [Accessed 5 January 2021].

Munhwa Broadcasting Corporation (1993) *MBC Newsdesk*, 19 September.

Munhwa Broadcasting Corporation (1998) *MBC Newsdesk*, 20 September.

NBC News (2014) 'North Korea to indict Americans Matthew Miller, Jeffrey Fowle', 29 June, Available from: https://www.nbcnews.com/news/world/north-korea-indict-americans-matthew-miller-jeffrey-fowle-n144021 [Accessed 5 January 2021].

North Korean Economy Watch (2012) 'KCTV changes evening news format', 12 March, Available from: http://www.nkeconwatch.com/2012/03/12/kctv-changes-evening-news-format/ [Accessed 5 January 2021].

Park, H. (2015) *The Capitalist Unconscious: From Korean Unification to Transnational Korea*, New York: Columbia University Press

Park, H. and Park, K. (1990) *China and North Korea: Politics of Integration and Modernization*, Vancouver: Asian Research Service.

Radio Free Asia (2014) 'Puk hallyut'ŭkchipp hallyu, chuminŭi salmŭl pakkunda' [North Korea Korean Wave special: Korean Wave, changing the lives of North Korean residents], 8 January, Available from: https://www.rfa.org/korean/weekly_program/radio_culture/fe-ck-01082014101810.html [Accessed 5 January 2021].

Riechmann, D. and Pennington, M. (2017) 'Estimates of North Korea's nuclear weapons are difficult to nail down', *Associated Press*, 18 August, Available from: https://apnews.com/article/53076b0dc7644f94b27 51134a1d9d76b [Accessed 5 January 2021].

Seoul Shinmun (2012) 'Puk yŏdaesaengdŭl, hatp'aench'ŭ yuhaenghanŭn kkadak algoboni...' [Reason behind popularity of 'hot pants trend' among North Korean female college students, as it turns out...], 4 May, Available from: http://seoul.co.kr/news/newsView.php?id=20120504500019 [Accessed 5 January 2021].

Seoul Shinmun (2018) 'T'albukcharamyŏn modu pwattanŭn tŭrama ch'ŏn'gugŭi kyedan' ["Stairway to Heaven", the drama that every defector has seen], 11 April, Available from: https://www.seoul.co.kr/news/newsView.php?id=20180411500029 [Accessed 5 January 2021].

Shim, E. (2015a) 'North Korea denounces US, Japan, South Korea for "cultural decadence"', *UPI*, 24 July, Available from: https://www.upi.com/Top_News/World-News/2015/07/24/North-Korea-denounces-US-Japan-South-Korea-for-cultural-decadence/8211437752312/ [Accessed 5 January 2021].

Shim, E. (2015b) 'Kim Jong Un declares "semi-state of war" after exchange of fire', *UPI*, 21 August, Available from: https://www.upi.com/Top_News/World-News/2015/08/21/Kim-Jong-Un-declares-semi-state-of-war-after-exchange-of-fire/7941440164397/ [Accessed 14 March 2021].

Shim, E. (2015c) 'Defector flew drones carrying flash drives into North Korea', *UPI*, 11 December, Available from: https://www.upi.com/Top_News/World-News/2015/12/11/Defector-flew-drones-carrying-flash-drives-into-North-Korea/1501449850667/ [Accessed 5 January 2021].

Shim, E. (2016a) 'Homer's nuclear incompetence has North Korea banning "The Simpsons"', *UPI*, 18 August, Available from: https://www.upi.com/Top_News/World-News/2016/08/18/Homers-nuclear-incompetence-has-North-Korea-banning-The-Simpsons/2841471537725/ [Accessed 5 January 2021].

Shim, E. (2016b) 'Kim Jong Un hails North Korea launch as "success of all successes"', *UPI*, 24 August, Available from: hhttps://www.upi.com/Top_News/World-News/2016/08/24/Kim-Jong-Un-hails-North-Korea-launch-as-success-of-all-successes/8641472087378/ [Accessed 5 January 2021].

Shim, E. (2017a) 'How North Korean missiles are helping to mask Kim Jong-un's fears', *South China Morning Post*, 26 April, Available from: https://www.scmp.com/comment/insight-opinion/article/2090482/how-north-korean-missiles-are-helping-mask-kim-jong-uns [Accessed 5 January 2021].

Shim, E. (2017b) 'Expert: Guam image in Kim Jong Un photo from 6 years ago', *UPI*, 17 August, Available from: https://www.upi.com/Top_News/World-News/2017/08/17/Expert-Guam-image-in-Kim-Jong-Un-photo-from-6-years-ago/8441502945528/ [Accessed 5 January 2021].

Shim, E. (2017c) 'North Korea defector: Hundreds of "reformers" killed after Jang Song Thaek purge', *UPI*, 17 October, Available from: https://www.upi.com/Top_News/World-News/2017/10/17/North-Korea-defector-Hundreds-of-reformers-killed-after-Jang-Song-Thaek-purge/6181508242576/ [Accessed 5 January 2021].

Shim, E. (2018a) 'North Korea workers returning to China, sources say', *UPI*, 5 April, Available from: https://www.upi.com/Top_News/World-News/2018/04/05/North-Korea-workers-returning-to-China-sources-say/5931522939234/ [Accessed 5 January 2021].

Shim, E. (2018b) 'Report: K-pop dance moves popular among young North Koreans', *UPI*, 11 October, Available from: https://www.upi.com/Top_News/World-News/2018/10/11/Report-K-pop-dance-moves-popular-among-young-North-Koreans/6721539273087/ [Accessed 5 January 2021].

Shim, E. (2019) 'Defector Kim Jong Un harsher than father, grandfather', *UPI*, 21 March, Available from: https://www.upi.com/Top_News/World-News/2019/03/21/Defector-Kim-Jong-Un-harsher-than-father-grandfather/7171553174052/ [Accessed 5 January 2021].

Spangler, T. (2014) 'Pirates swarm over "The Interview" after Sony's digital release', *Variety*, 25 December, Available from: https://variety.com/2014/digital/news/pirates-swarm-over-the-interview-after-sonys-digital-release-1201387196/ [Accessed 5 January 2021].

TV Chosun (2013) 'Puk kajŏngjip naebu yŏngsang, nuga ch'waryŏnghaenna' [Footage of a North Korean home], 3 February, Available from: https://www.youtube.com/watch?v=SQ595ozU09w [Accessed 5 January 2021].

VOA Korea (2013) 'Puk'ansŏ han'guk tŭrama param...ch'oegŭn tansok kanghwa [Winds of South Korean drama blowing in the North ... Crackdown strengthens in recent times]. 22 November, Available from: https://www.voakorea.com/a/1795062.html [Accessed 5 January 2021].

Yonhap TV (2018) '"Mogijangeul dandanhi chyeoya"...pukhan, jabonjuuimunhwa chimtu gyeonggye' ['Mosquito net must be firmly in place'... North Korea warns of invasion of capitalist culture], 28 August, Available from: https://www.youtube.com/watch?v=2qT7F1Z2jQQ [Accessed 5 January 2021].

YTN (2013) 'USB ro puk'an nae namhan tŭrama hwaksan' [South Korean dramas in North Korea spreading via USB], 22 October, Available from: https://www.ytn.co.kr/_ln/0101_201310221702578310 [Accessed 5 January 2021].

Techno-identity and Digital Labour Condition

Introduction to Part III

Peichi Chung

This part brings together four book chapters to discuss how techno-identities and digital labour conditions in the Northeast Asian region are shaped in the domains of esports, robotics, and pachinko. In Chapter 7, Keung Yoon Bae examines the transnational labour flow of South Korean esports. Bae uses Nick Dyer-Witheford's concept of biopower to examine the global participation of South Korean esports despite restrictions on intellectual property rights controlled by international game developers. In Chapter 8, Peichi Chung adopts a political-economic perspective to evaluate new labour conditions in South Korea's esports. Keiji Amano and Geoffrey Rockwell focus in Chapter 9 on the historical turn to understand Japanese cinematic imageries associated with pachinko, to show how the game – as a racialised technology – reflects Japan's deeper intercultural struggle among subcultural groups. In the final chapter, Shawn Bender researches the emerging human–machine relationship in Japan's robot therapy developed to serve the country's ageing society by functioning as feeling machines designed to increase the sociality of Japanese elders. The four chapters collectively reveal diverse techno-cultural spaces that speak about the existence of various subcultural and emerging class subjectivities in Japan and South Korea.

Keung Yoon Bae studies the global competitiveness of South Korean esports players, using Nick Dyer-Witheford's concept of biopower to explain the global circulation of South Korean esports players. The chapter first describes the commercial operation of esports organisations, which apply business principles to professionalise esports. This explains the desire of game developers to seek control over their players through intellectual property rights management. Bae sees corporations' intention to build governance across the global esports landscape as a reflection of militarised hypercapitalism. The second part of the chapter elaborates on South Korea's use of technology to disrupt corporation control. Bae perceives that local features embedded in South Korean culture are contributing factors in South Korean competitiveness. It is common to see South Korean esports players joining US or European teams, and Bae sees that the key reason behind the global circulation of South Korean esports players is that these players tend to employ a more dedicated attitude to training, self-improvement, and mutual respect for peers and elders, which makes South Korean players more team-oriented. Another unique local feature, the PC *bang*, lies at the root of the gaming culture that has contributed to the biopolitical production of South Korean professional esports. The Korean PC gaming infrastructure offers its esports players advanced technologies to help them gain the competitive advantages necessary to enter the global professional scene.

Peichi Chung focuses on the analysis of player power in creating new labour conditions in South Korea's networked economy. She examines the role of South Korean esports players in restructuring power relations among Northeast Asian countries in digital gaming. Chung analyses the political-economic context of the country's esports industry. The chapter focuses on player welfare as one of the main concerns in shaping the public ecosystem of South Korean esports. It outlines the contribution of the Korean Esports Association and International Esports Federation to advancing the player awareness policy agenda in both South Korean and global esports policies. Chung concludes with a section on amateur esports players who have demonstrably contributed to building a competitive esports industry by creating employment of their own, which helps to extend the country's esports ecosystem beyond the practice and control of global game developers.

In Chapter 9, Keiji Amano and Geoffrey Rockwell study the representation of play in selected films about pachinko in Japan. The chapter examines the process by which pachinko turns into a metaphor for Japan through a series of Japanese and Western movies reviewed in chronological order from the 1950s to 1990s. In their analysis, Amano and Rockwell discern such imageries in cinematic representations of pachinko as a 'harmless pastime of youth', a 'playful randomness of life', 'living places for the ethnic Japanese-Korean community', 'opaque black box profiting from anti-social forces' or a 'dark depiction of Japan's invisible caste system in racial and class politics'. Pachinko's transition from the early films of the 1950s to the later films of the 1990s reveals the existence of various techno-cultural spheres of contentious racial politics in Japanese society. This chapter demonstrates a profound case of pachinko play in connection with identity transformation of Japan's important Korean ethnic minority.

The last chapter of the book presents a new type of physical and affective labour in Japan's ageing society. Here, Shawn Bender offers an ethnographic account of the new phenomenon of robot–human co-existence in the country's elder care service. As Japan now faces growing demand from its ageing population, robotic technology has developed an innovative trend that can alleviate the needs of labour and care service productivity. Bender observes that these robots are machines that have been programmed to feel and be felt. For example, robotic tools can provide dementia patients with simulated experiences that remind them to follow through with everyday rituals of waking, bathing, eating, and socialising activities. As robotics machines provide a new kind of infrastructure for engagement, robot therapy succeeds in offering a platform for sociality that helps professional teams of robot therapists expand their care to more nursing homes and other locations.

Three issues emerge from these chapters in relation to techno-identities and digital labour in Japan and South Korea. First, the continuing disparity of labour conditions in the three countries studied are embedded in different technological developments (see also Introduction about esports labour). A current divergence is prevalent along longstanding lines of political division reflected the greater Cold War structure, dividing communist North Korea from Western capitalist Japan and South Korea (see also Introduction about the impact of the Cold War on contemporary geopolitics in Northeast Asia). Fans of Korean Wave in North Korea are digital labour working lower technological forms, secretly consuming South Korean media content through small digital devices like DVDs and USBs (Zhang and Lee, 2019; see also Chapter 5). The digital labour conditions in Japan and South Korea reveal a labour transformation in progress in the new realms of human–machine interactions. For Japan, robotic technology introduces a new kind of labour productivity through robot therapy. For South Korea, gaming skills honed by South Korean esports players point to a new type of labour transformation based upon flexible specialisation in the human-to-system context. In general, the machine has restructured Japan's workplace to incorporate non-human robots with human labourers. The global circulation of South Korean esports labour shows us the speedy mobility of the digital labour flows produced. Given the extent of technological progress in the global economy, it remains important to engage in labour politics that will help ensure equal opportunity for class mobility in the digital capitalist economy.

The second issue relating to techno-identities and digital labour is evident in the blurring corporation–labour division achieved through automation. Because automation always concerns work efficiency, labour productivity, and job performance, the chapters on Japan and South Korea compel us to ask, does automation really solve the problems of poverty and inequality? Does robotic therapy eventually address the needs of patients with dementia? What about the limits of robot care given the lack of communication with elders, which Bender brings up at the end of his chapter? This is a problem not only for Japan but also for South Korea, whose population is currently ageing at the fastest rate among Organisation for Economic Co-operation and Development members (Kim, 2019). As society undergoes workplace restructuring through computing technology, one also needs to question the existence of worker autonomy. How much does esports professionalisation change new class entry into waged employment? How does automation in the form of esports enable professional players to access their mode of production when they generate value (or data) in today's networked content creation process?

The third issue lies in seeing pachinko as techno-cultural sphere of complex ethnic identity struggle among *Zainichi* (Koreans residing in Japan). Pachinko is a large pinball gambling business that emerged as leisure entertainment in the 1930s. The industry took off after World War II and became a popular adult pastime in Japanese society. Japan's pachinko business is predominantly dominated by Korean-Japanese people. Some of the pachinko parlours are controlled by North Korean agents who migrated to Japan from North Korea during the Japanese colonial period. Because of their favourable political perspective towards North Korea, some of the pachinko parlour owners have sent millions of dollars to North Korea. In 2018, Pachinko was one of Japan's largest entertainment markets, earning more than 30 times more profit than the casino industry in Las Vegas (Chan, 2018). The pachinko industry employs more than 300,000 people. It provides a better livelihood for many Korean-Japanese people, since opportunities for other businesses are blocked due to citizenship problems Koreans face in Japan. In all, pachinko is a contested cultural location which not only brings entertainment to Japan's parlour customers, but also symbolises the influence of play technology that preserves culture and thereby ensures perseverance among ethnic groups that fight for visibility in Japanese society.

The development of advanced media technologies in Northeast Asia covered by the four chapters reveals how ongoing entangling international conflicts among North Korea, Japan, and South Korea continue in the digital domain. Political competition and negotiation in the techno-cultural spheres of pachinko, robotics, and esports are expressed in ways that enunciate a unique Northeast Asian regionalism. The perseverance of *Zainichi* subjectivity is one of our cases that sheds light on people's use of technology to uncover the immigrant politics in Japan's ethnically homogeneous society. Pachinko, as we look into the racial domain of this digital technology, transforms various stories of subculture into valuable tales for immigrant cultural resistance and worker autonomy (Lee, 2017).

References

Chan, T. (2018) 'Japan's pinball gambling industry rakes in 30 times more cash than Las Vegas casinos', *Business Insider*, 26 July, Available from: https://www.businessinsider.com/what-is-pachinko-gambling-japan-2018-7 [Accessed 11 January 2021].

Kim, S. (2019) 'South Korea's robots are both friends and job killers', *Bloomberg Businessweek*, 11 November, Available from: https://www.bloomberg.com/graphics/2019-new-economy-drivers-and-disrupters/south-korea.html [Accessed 11 January 2021].

Lee, M. (2017) *Pachinko*, London: Head of Zeus.

Zhang, W. and Lee, M. (2019) 'Black markets, red states: media piracy in China and the Korean Wave in North Korea', in Y. Kim (ed) *South Korean Popular Culture and North Korea*, London: Routledge, pp 83–95.

7

'Too Many Koreans': Esports Biopower and South Korean Gaming Infrastructure

Keung Yoon Bae

On 10 January 2018, the inaugural season of the Overwatch League (also commonly abbreviated to OWL) began. As a major professional esports league for the first-person shooter game *Overwatch* by Blizzard Entertainment, OWL was not the first large-scale, global, franchised esports league. But it was doing something unique that would set it apart from other esports leagues for games such as *League of Legends*, *Dota 2*, and *Call of Duty*: It was attempting a geolocation model in esports. Unlike other esports leagues with endemic esports organisations that had few, if any, city-based ties, OWL would feature city-based teams, with branding and 'identities' tied to different cities and attempts to create local fanbases in those cities. While the first two seasons of the OWL would be held at the Blizzard Arena in Burbank, Calif., from the third season onwards the league would attempt to institute a new system of 'home' and 'away' matches, with esports events being hosted in the teams' home cities.

With the announcement of a geolocation model franchising system, endemic esports organisations (which are often founded and owned by people involved in esports, such as former professional players) that had fielded teams in *Overwatch* esports either quickly folded their operations or bought a city-based franchise spot for US$20 million. Examples of the former are FaZe Clan, Luminosity Gaming, and Splyce; the latter include Cloud9 and Immortals. Among the first twelve teams to join the league in its inaugural season in 2018 were the Los Angeles Valiant (owned by the US-based esports organisation Immortals), the London Spitfire (owned by the US-based esports organisation Cloud9), the New York Excelsior (owned by Sterling VC, which also owns the Major League Baseball team the New York Mets), and the Boston Uprising (owned by Robert Kraft of the Kraft Group, which also owns the popular American football team the New England Patriots). But, as was quickly noted by viewers and fans, the emphasis on cities did not translate into local players representing their cities and teams. As an ESPN esports journalist noted, this was due in part to 'the Overwatch League's insistence on having a completely open league' (Rand, 2018). That is, it would not require teams to hire players from the regions that the teams represented (also known as region locking).[1] One result is that some teams ended up fielding rosters made up entirely of South Korean nationals.

Why did an American esports organisation field an entirely South Korean roster of players in order to represent London in a US corporation-run 'global' esports league? Or, more crudely (and politically incorrectly) put, why are there so many Koreans in esports? What are the various socio–economic and geopolitical forces involved in creating such circumstances in competitive video gaming? How does an examination of South Korean esports contribute to an understanding of the intersection of work and play in contemporary gaming and streaming culture, immaterial labour, and global dynamics of corporate control? These are some of the questions asked in this chapter while certain facets of global esports developed in the 2010s will be introduced. As the introductory example of OWL shows, the competitive scene in *Overwatch* showcases not only the ambitious scale on which game developer companies are envisioning esports' future, but also the myriad contradictions that shape esports at this juncture, illuminating issues of infrastructure, labour supply, and corporate control in sports.

By looking at OWL, this chapter adds to a discussion on esports professionalisation and industry relations, which have already been discussed by T.L. Taylor in *Raising the Stakes: E-sports and the Professionalization of Computer Gaming* (2012) and Dal Yong Jin in *Korea's Online Gaming Empire* (2010). I will be discussing these topics primarily in a South Korean context to enrich the conversation, but will also demonstrate that examining esports in an exclusively regional context is no longer realistic or even sensible in today's landscape, as evidenced by the example of the London Spitfire. In other words, South Korean esports exists in deep connection with other esports scenes in the US, China, Europe, Japan, and so on, not merely in terms of players connecting with each other at international competitions, but also in terms of labour flow, power dynamics, and racialised and nationalised identities.

In the first section, I introduce the landscape of esports and the significant changes it has undergone in the past decade. In the second section, I take cues from Dyer-Witheford and de Peuter (2009) to apply the framework of biopower to gaming and esports, in particular the ways in which game developer companies seek to retain control over not only the gaming populace, but also the sport of gaming. Dyer-Witheford and de Peuter, using the company Blizzard and its control over the Massively Multi-player Online (MMO) game *World of Warcraft* as a case study, use biopower as a key framework to encapsulate the degree to which video games corporations genuinely 'rule' the digital realms that they have created, and how by extension they exercise a similar regulatory power over real-life, out-of-game activities. I propose that this framework could also be applied to esports, in that the video game companies exert a similar level of absolute control on the competitive sports of their games by way of copyright law, and in doing so they fundamentally define not only in-game activities but also the shape of the competitive landscape and the associated economy of esports. Finally, I illuminate how regionally specific factors, such as local infrastructure, come into play to disrupt or complicate that control, despite the lengths to which companies go to exert it.

I must also acknowledge that the scope of this chapter is by no means an exhaustive or comprehensive look at esports: For the purposes of my analysis, I draw heavily from two major professional esports leagues, the OWL based in Los Angeles and the League of Legends Championship Korea (LCK) based in South Korea, both of which are PC-based (rather than console-based) esports and are directly administered by their respective developer companies, Blizzard Entertainment and Riot Games. As Taylor (2012), Witkowski (2012), and other scholars have covered in their own works, many other professional esports leagues are not PC-based, such as *Call of Duty* and *Halo*. Moreover, many esports – especially fighting games, arcade games, and sports games – are also not PC-based (Peterson, 2018). However, this chapter is more interested in locating the global, biopower dimensions of esports in a time of increasing corporatisation, where esports leagues are transitioning into multi-million dollar franchising models, and *League of Legends* and *Overwatch* are two games that exemplify this transition.

What is esports? A brief introduction to the landscape

I first introduce basic systems and practices that have been established in esports since 2000, before delving into the nitty-gritty of recent development. Esports, or professional competitive video gaming, has a history before its current, somewhat PC-focused landscape. Taylor (2012) points out that as early as the 1970s, video game players competed against each other in gameplay, with arcade machines making available high score notations, and arcades being fashioned as a space for human-versus-human (as opposed to human-versus-machine) competition (Taylor, 2012). In the late 1990s, after the fall of arcades and the rise of networked PC gaming, esports experienced another boom through popular multi-player games such as *Quake* and *StarCraft*, though plenty of game enthusiasts will attest that the Fighting Game Community (FGC) – which primarily plays arcade and console-based 1-versus-1 fighting games – was also active and vibrant throughout the 1990s and maintains a strong presence to this day (Cuevo, 2019).

Though one may assume that network technologies would have made esports a primarily online activity, esports has actually always been defined by physical gatherings, where players and spectators gather in convention halls, arenas, and massive stadiums to watch the gameplay unfold on screens. With the rise of online streaming, platforms such as *Twitch* also broadcast events live. The importance of a physical location came from the 1990s gaming community that often held 'LAN (local area network) parties' to compete with each other. Gamers gathered in the same physical space to connect their devices to each other and play, but today the set-up has more to do with reducing latency for players (so that their commands go through smoothly to the game server without any lag due to a slow Internet connection) and staging spectacular events for broadcast.

Since the 2000s, esports professionals have become increasingly affiliated with esports organisations ('orgs') that essentially act as professional agencies for the players. 'Endemic' organisations, such as FaZe Clan and Splyce, not only act as player agencies, but also manage sponsorships, host esports events, and create and sell merchandise. Some teams receive sponsorships from multiple sources. For example, US-based Team Liquid lists companies such as Alienware, Monster Energy, and SAP among their partners (Team Liquid, 2019). Others are primarily funded by and named for major corporations, such as the LCK teams SKT T1, KT Rolster, and Jin Air Greenwings. Sponsorships, which are prominently advertised on players' live streams, and streaming revenue are the primary income sources for esports organisations (Pannekeet, 2019).

With such agencies 'managing' the players and the teams, 'team houses' have become a standard practice for team-based esports like *League of Legends* and *Overwatch*. The team house originates from *StarCraft* practices in South Korea:

> Most teams have their own dormitories and practice rooms in Seoul, where most professional games take place, and pro gamers live in their team's dormitory. [...] Professional teams support players by providing dormitories and practice rooms because some players come from areas with no money, and the boarding house system makes it easy for team managers and companies to manage team training and scheduling. (Jin, 2010: 91)

While the practice of players cohabiting may have initially begun as a cost-cutting measure, it is also notable that *hapsuk hullyŏn* (bootcamp training) has a long history in elite athlete training in South Korea, especially in the 1970s, when the government used it to boost the country's performances at international sporting events. In addition, *hapsuk* is also instituted in K-pop idol training.

Esports is in many ways the epitome of the intersection of work and play, where 'play' has literally become a professional occupation, with an entire industry formed around it. And, as is clear from this brief overview, the industry is comprised of multiple entities: players, fans, game developers, esports organisations, and streaming platforms. They all play a part in shaping esports practices and systems, with some entities often taking on multiple roles at once. For example, *Afreeca* and *Bilibili*, two popular streaming platforms in South Korea and China respectively, also run their own esports organisations. This landscape, however, is rapidly changing, with recent shifts in *Overwatch* and *League of Legends*, which I will examine in this chapter, specifically in terms of game developer control.

It bears mentioning that an ongoing debate in esports is about whether it is a real occupation or sport in the same way that professional baseball, football, or golf are acknowledged to be. This is admittedly an existential question with which esports itself is often obsessed; as Emma Witkowski pointed out in her article on *Counter-Strike* LAN tournaments, 'physicality has unquestionably been the Achilles heel of e-sports in terms of its sporting legitimacy' (Witkowski, 2012: 356). For the purposes of this chapter, however, I will be approaching esports as a legitimate sporting occupation and professional players as athletes. While there are indeed distinct characteristics that set esports apart from traditional sports – for example, esports has the predominance of online streaming, which enables much more direct engagement between fans and players – many of esports' fan practices, live events, and broadcasts look a great deal like mainstream professional sports already, and will likely continue to become even more so as esports grows in prominence.

Franchising, professionalisation, and control in esports

Having briefly covered the basic concept of esports and its relatively short history, I will now summarise the more recent development in esports through the examples of the OWL and *League of Legends* esports. These changes come at a time when there is increasing recognition of the size and potential of the esports market, as evidenced by the 2019 *Newzoo* financial report (Pannekeet, 2019). While I must note that these market numbers, which declare that the esports market passed the billion dollar mark in 2019, may not tell a complete story because they were put forth by financial analysts and/or game companies (D'Anastasio, 2019), the increased attention to the esports market goes hand-in-hand with the increasing 'vertical integration' between game development and esports, with the developer companies wielding a much larger influence than ever before.

Some TV viewers in the US were startled in early 2019 to see video game competitions being aired on mainstream television channels, along with football and baseball. In February, the Overwatch League announced that some of their broadcast matches marked 'the first esports competition to air live on ESPN's flagship network in primetime, and […] the first broadcast of an esports championship on ABC' (Fogel, 2018). Developments like these are part of the recent attempts in esports, which had in the previous decades relied on third-party competitions and grassroots-organised tournaments, to move into a more corporate-directed, mainstream-style franchising model in the mode of the National Football League (NFL) and National Basketball Association (NBA). Such a shift is not surprising for those who have been observing the increased interest and rising financial stakes in esports in the 2010s: Since 2014, the total prize pool for *Dota 2*'s 'The International' tournament (the biggest and most important tournament for the esport, which combines a base prize pool of US$1.6 million and a crowdfunded prize pool) has surpassed US$10 million, with its most recent 2018 prize pool reaching a staggering US$25 million (Rodriguez, 2019). In 2018, Riot Games reported that their 2018 *League of Legends* World Championship Finals had attracted 99.6 million unique viewers worldwide (though it is difficult to independently confirm these numbers, as Chinese viewership numbers are difficult to gauge accurately), 'broadcast in 19 different languages across more than 30 platforms and television channels' (*League of Legends*, 2018).

The size and spending capacity of esports audiences rival those of traditional sports and have attracted pre-existing, traditional sports organisations and investors to get involved. In 2018 both the North American *League of Legends* Champion Series (LCS hereafter) and the OWL introduced official franchising systems for their respective leagues. While the LCS evaluated and accepted bids from investors and esports organisations based on their financial capacities, Blizzard attempted something different for the OWL: 'the company wanted to take what it saw as the best aspects of traditional sports – namely, permanent franchises, professional ownership groups, and teams based in specific cities – and merge them with the burgeoning world of esports' (Webster, 2018). The basic rationale behind a franchising system is that it provides financial stability not only for teams and fanbases, but also for players' careers and their financial security.[2]

However, when viewing these developments in the context of a slightly longer history of esports and intellectual property disputes, the franchising system is but another step in a longer trajectory of game developer/publishing companies cementing their control over every aspect and extrapolations of the game. Unlike the NFL or the NBA, which do not actually own the sports of football or baseball, Riot Games and Blizzard own their respective games and, by extension, the esports that can be born from those games. Part of this heightened need for control can be traced back to Blizzard Entertainment's history of contested ownership over esports in South Korea, which I will explain briefly.

The success of Blizzard's iconic RTS (real-time strategy) game *StarCraft* (and its expansion, *Brood War*) in South Korea's PC *bangs* (Internet cafés) has been well-documented in both journalistic and academic works, such as Jin's *Korea's Online Gaming Empire* (2010), McCrea's 'Watching *StarCraft*, strategy, and South Korea' (2009) and Huhh's 'The "Bang" where Korean online gaming began: the culture and business of the *PC bang* in Korea' (2009). Blizzard's experience in South Korea with regards to the StarCraft 1 esports scene ultimately shaped its approach toward controlling esports. According to an in-depth *Variety* article by Will Partin (2018), in which he interviewed

Blizzard's foundational figures such as Mike Morhaime and Patrick Wyatt, he raised questions about an entire 'afterlife' of the game that the developers had not considered:

> What should Blizzard's role be in fostering this nascent esports scene? Should the game's designers update the game when unfairly strong strategies emerged, or wait for players to figure something out on their own? Did the company have a responsibility – legal, ethical, or otherwise – to support those who wanted to turn 'StarCraft' into a career? And, above all, how was Blizzard going to monetise the cultural vein they'd struck? If you were one of 'StarCraft's' designers, after all, you probably saw the game's professional scene as one sign of a job well done. But if you were an investor, you saw value being created, and lost. (Partin, 2018)

As Partin described it, Blizzard 'played a marginal role in shaping its esports scene in South Korea' for *StarCraft: Brood War*, with the reins mostly being held by the Korean eSports Association (KeSPA) when it came to governing and organising esports competitions, as international trade agreements and intellectual property rights had not yet caught up to phenomena such as esports. As legal scholar Burk (2013) asserted in his article on Blizzard and KeSPA's disputes over esports, their conflict over *StarCraft* was a key case in determining the question of intellectual property ownership in esports. Burk wrote:

> Ownership of casual game performances is largely a moot point [...] because there is usually not enough at stake for anyone to seriously challenge it. Players are often [...] attached to their online depictions, but that attachment is seldom monetised [...] The landscape changes when the performance has demonstrable worth because the human behind the avatar is a professional player, who attracts the attention of fans, advertisers, and sponsors; and who generates revenue for his team and his league by means of his performances. (Burk, 2013: 1545)

Blizzard learnt their lesson and used an entirely different approach with their much-anticipated follow-up, *StarCraft II*. While Paul Della Bitta (then senior director of Global Community Development and eSports at Blizzard), in his communications with T.L. Taylor, 'emphasized Blizzard's support for the e-sports community' and 'pulled back from painting too strong a picture of Blizzard being very hands-on in relation to third-party e-sports events',[3] even going so far as to say 'we're not in the business of e-sports' (Taylor, 2012: 163), Will Partin's observations about Blizzard's different approach for *StarCraft II* speak otherwise. Partin pointed out that Blizzard eliminated LAN support for the sequel and routed it through the company's servers. This would give them 'unprecedented control over the game and an inherent advantage in negotiations with potential broadcast partners' (Partin, 2018). He also noted the new free trade agreement finalised between the US and South Korea in 2007 'affording multinational companies new legal protections, including for intellectual property', and the new user policies that shifted to an end-user license agreement, which solidified Blizzard's intellectual ownership over the game (Partin, 2018). These shifts have remained concretely in place ever since, and have squarely put the reins in the hands of Blizzard in shaping whatever esports scene arose around their games. This status quo also applies to other developers, like Riot Games (*League of Legends*) and Valve (*Counter-Strike*), and such levels of control are now considered standard: 'Riot is the final arbiter of what's fair and right in their leagues. With the LCS, where all the baseline salaries are guaranteed by Riot and the league is under their control, Riot's word is law' (Zacny, 2016).

The degree of control that the game companies hold over esports has not gone unquestioned by the owners of esports organisations who have sought ways to share the revenue with the game companies, in the way that traditional sports clubs have some broadcasting rights that translate into a steady revenue stream. In an interview, the Chief Gaming Officer of Fnatic (a major European esports org), Patrik Sättermon, pointed out the skewed power dynamic in esports as compared to traditional sports: 'it's a significant difference between football, for example, and *League of Legends*: in *League of Legends* you have someone with ultimate power, the people who constructed and developed that game – that's on a perpetual basis' (*The Versed*, 2017). While Riot Games has introduced some initiatives for revenue sharing with its *League of Legends* teams, it has yet to release its exclusive grip on broadcasting rights. In 2010, KeSPA challenged Blizzard in court over the broadcasting rights for *StarCraft*: 'KeSPA had been broadcasting *StarCraft* for years at this point, and argued they helped make *StarCraft* what it is. Blizzard argued it was their game, and they had the rights to decide who could broadcast the game' (Ellis, 2016). The dispute did not last long: 'it was a very open and shut case in that Blizzard owned the IP and KeSPA pretty much caved on all accounts' (Ellis, 2016).

It is not difficult to see the points made by esports organisations: Esports is a collectively created phenomenon, and applying the standard of intellectual property is simplistic and unfair to all the parties involved in esports. Unlike mass media content, major esports like *League of Legends*, *Overwatch*, and *StarCraft* owe a great deal to gaming scenes located outside the United States. *StarCraft* esports, and arguably modern esports itself, was born in Korean PC *bangs*, and *League of Legends'* almost indomitable popularity in China is part of the power it wields as one of the world's biggest esports. Before the OWL launched, the biggest and most-watched esports tournament for *Overwatch* was the APEX series, a tournament organised and broadcast by OGN, a major Korean gaming channel. The formerly localised, collectively constructed, multi-party nature of esports has increasingly been replaced with unilateral developer oversight, a shift that sways the centre of control from the various regions of the world over to the US.[4] However, I will examine in the next section how this developer company control over esports has faced significant disruption due to regionally specific technological conditions.

Esports, biopower, and biopolitical production

In *Games of Empire: Global Capitalism and Video Games* (2009), Dyer-Witheford and de Peuter present video games as both exemplary products and constitutive elements of the concept of 'empire' as put forth by Michael Hardt and Antonio Negri in *Empire* (2000): 'planetary, militarized hypercapitalism' (Dyer-Witheford and de Peuter, 2009: 19). They argue that 'virtual games are media constitutive of twenty-first century global hypercapitalism, and, perhaps, also lines of exodus from it', as originally products of the US military-industrial complex. Virtual video games are also deeply enmeshed in 'methods of accumulation based on intellectual property rights, cognitive exploitation, cultural hybridization, transcontinentally subcontracted dirty work, and world-marketed commodities' (Dyer-Witheford and de Peuter, 2009: 39–40). Dyer-Witheford and de Peuter locate video games within a 'fractured economic order', where both consumption and production has been globalised, albeit in an uneven and skewed fashion (Dyer-Witheford and de Peuter, 2009: 23). They further look to geographically located/shaped modes of biopower and conflict, namely intensive gold farming operations in China that sell in-game currency and goods for offline world currency to US- and Europe-based players, which threatens the publishing company's all-encompassing grasp over the in-game economy and value creation capabilities. As in-game currency, like gold in *World of Warcraft* (*WoW*), is essential in many games to play at higher levels with better 'equipment' like armour and weapons, it follows that some players are willing to pay for services that would help them acquire gold faster and easier. In the in-game market economy, millions of Chinese *WoW* players use the game to trade in-game gold for real-world cash,[5] which directly goes against Blizzard's terms of use for the game: 'No one has the right to "sell" Blizzard Entertainment's content, except Blizzard Entertainment' (Dyer-Witheford and de Peuter, 2009: 243).

From the Foucauldian lens of biopower, these exchanges illuminate the friction within MMO games in which biopower from the developer companies struggles to control the 'biopolitical production' activity of the masses (Dyer-Witheford and de Peuter, 2009: 223). The authors expand the original Foucauldian definition of biopower, referring to government regimes' power and discipline over life and death itself, in order to explain how a game developer administers and governs at a *double* level: 'virtual and actual biopower, proceeding simultaneously at two levels – that of the in-game digital world, with its enchanted territories, heroic characters, and fearsome monsters, and the real-life apparatus of shard servers, system administrators, and fee-paying subscribers' (Dyer-Witheford and de Peuter, 2009: 223). Just as Foucault describes biopower as the administering of life (and death) through careful control and regulation, Dyer-Witheford and de Peuter view Blizzard's tight control over *World of Warcraft* gameplay as exemplifying 'the enlarged scope of twenty-first century biopower' (Dyer-Witheford and de Peuter, 2009: 223). They argue that the predicament of gold farmers in *WoW* is both a challenge to the 'publishers' monopoly over the value-producing 'play-bour"' (Dyer-Witheford and de Peuter, 2009: 225) of World of Warcraft as well as an indicator of Marx's concept of 'subsumption', in which 'fully Foucauldian regimes of biopower' are carried into virtual game spaces (Dyer-Witheford and de Peuter, 2009: 257–8). Furthermore, their analysis of Blizzard's approach towards gold farming in *WoW* presents a framework that could be applied to esports, especially as the game developer's attempts to maintain an iron grasp on every step of value creation resonates strongly with the way in which Blizzard took over the *StarCraft* esports scene once they 'saw value being created, and lost' (Partin, 2018).

The demographic of South Korean esports players I discuss in this chapter cannot be said to work under conditions comparable to the gold farmers of Azeroth/China in *World of Warcraft*, nor do they exist in a similarly polarised state of biopower/production. However, the framework of biopower nonetheless is apt to apply, in that fundamentally esports is, like gold farming, a form of value creation that has sprung outward from in-game activity to become an out-of-game value creation system that encompasses not only the 'play-bour' of the players themselves but also the broadcasting company using the game screen on their broadcast, the esports observers operating in-game 'cameras' to show the action, and the staff organising the teams in the game lobby and making sure no cheats or hacks are running on their computers. Just as Blizzard sought to maintain their 'monopoly over the value-producing "play-bour"' in *World of Warcraft* (Partin, 2018), they established firm control over the value creation chain of esports for their games by asserting their IP rights. The recent manoeuvres from Blizzard and Riot to 'legitimise' their esports into franchised sports leagues are a continuation of that trajectory. The companies' institution of formal, globalised leagues – the OWL with its 20 represented cities,[6] and *League of Legends* with its many different regions represented at the World Championships –[7] are attempts to impose a uniform, systematised order to the activity of competition. Rather than leaving the various regions to run their own tournaments autonomously, the companies have increasingly exerted pressure to centralise and vertically integrate esports across the globe, pushing out third-party competition organisers, broadcasters, and content creators.

The degree to which this monopoly has been maintained can be most clearly seen in the way that both Riot and Blizzard extricated themselves from their collaborative relationships with the South Korean cable television channel OGN (formerly known as ongamenet). OGN, a broadcasting company that rose to prominence with the 2000s *Starcraft: Brood War* scene in South Korea, has long been a mainstay in the Korean esports landscape, 'the standard that every other esports broadcasting company in Korea strives for' (Isaakov, 2016). OGN was involved in *League of Legends* competitions from early on, in 2012, and likewise produced and broadcast *Overwatch* competitions starting with their APEX series in 2016, soon after the game was released. While it is difficult to measure the success of an esports broadcast – there is still debate on how streaming viewership is tabulated and how it should be judged – it was widely agreed that OGN's broadcasts were well-produced and enjoyable, with APEX being called 'once one of the largest and most prestigious Overwatch tournaments in the world' (Carpenter, 2018). Regardless of what OGN could offer, however, both Riot and Blizzard opted to bring their video production and broadcasting operations in-house, fully vertically integrating their esports divisions and eliminating any reliance on third-party companies, despite the 'inherent advantage' that the publishers now had in negotiations with broadcasting partners. The companies have clearly demonstrated their need to eliminate the risk of esports becoming separately associated with a third party, even if it is within a single region, and ensure that esports is integrated into the developer company, even at the cost of the quality of the esports broadcast (which has been noted and criticised by fans and viewers, especially in the case of *Overwatch*).

Furthermore, the concept of a global league as put forth by Riot and Blizzard fulfils the promise and myth of esports, which has always been that it is a meritocracy. No matter who you are, what race, gender, or nationality, as long as you have the skill, you can become someone in esports. It's why OWL players such as Jay 'Sinatraa' Won and Pongphop 'Mickie' Rattanasangchod received so much attention in the league's inaugural season – they were star examples of young players who, based purely on their skills, found themselves competing in a global league and earning sizable amounts of cash (Wolf, 2017). Like the FIFA World Cup or the Olympics, the OWL and the *League of Legends* World Championships present themselves as a stage on which anyone in the world can compete, with Riot and Blizzard as the benevolent overseers. In terms of soccer, they are both FIFA and the game itself.

Which brings us back to the original question raised at the beginning of the chapter: Why is a US-based esports organisation, representing London in the league, fielding an all-Korean roster of players? Why, in the 2019 season of the OWL, were the London Spitfire, New York Excelsior, Toronto Defiant, Hangzhou Spark, Guangzhou Charge, Shanghai Dragons, and Florida Mayhem represented almost entirely (if not totally) by South Korean players? Why, in 2016, did Riot Games institute more stringent rules to their Interregional Movement Policy (which already imposed a 'region lock' for teams that would limit the number of non-resident, 'imported' players for each team) for *League of Legends* esports teams, requiring 'imported' players to spend longer in a specific region to be considered full resident players (LoLesports Staff, 2016), a rule change that immediately impacted 'nearly every team [...] outside of South Korea' (Volk, 2016)? And even after that rule change, why was Team Liquid, the winner of the LCS (*League of Legends*' North American league), represented by two South Koreans in their five-man squad? Why is Invictus Gaming, a Chinese gaming organisation owned by the scion of the Wanda Group that won the 2018 *League of Legends* World Championship, spearheaded by two-star Korean players? Simply put, why do these global leagues not look that, well, global?[8]

The short answer to all the above questions is that Korean esports players are indeed some of the best players in the game, a fact of esports that has birthed many a discussion thread on online esports forums, as well as countless articles. Esports discourse often subscribes to the idea that Korea's prominence in esports, and the seemingly higher skill level of Korean teams and players, comes from elements inherent in Korean culture: the idea that traditional Korean culture (and Confucianism) causes Korean players to have a more dedicated attitude to training and self-improvement, more respect for their peers and elders, and a tendency to be more team-oriented (rainha, 2015). The explanation serves to idealise and exoticise Korean players and their achievements, while also safely maintaining that ultimately esports is still functioning as a meritocracy. What the discourse in esports often lacks is a recognition that Korean players have unparalleled access to high-speed Internet and good gaming PCs at extremely affordable prices, in the form of ubiquitous PC *bangs* all over the country. In a 2019 interview, Lee 'Faker' Sang-hyeok – arguably the most decorated *League of Legends* player in the world – acknowledged that his family was on national welfare as he was growing up. 'But', he continued, 'that environment didn't impact me. I played games as much as I wanted, so I didn't think I was poor […] If I wanted to go to a PC *bang* I would just save up some bus fare' (Lee, 2019).

Part of the authority that developer/publishing companies like Blizzard and Riot assume in their governance of the esports realm comes from the idea that they can ensure a more meritocratic and improved ecosystem in their respective games and esports, by making the systems centralised and uniform. The loss of LAN functionality starting with *StarCraft II* was a classic example of this: 'Matches would have to be coordinated through a centralised server system, Blizzard's Battle.net, rather than be set up ad hoc at tournaments around the world' (Taylor, 2012: 162). Franchising – with its promise of standardised minimum salaries, training conditions, and player contracts – similarly appears to ensure a more meritocratic and level playing field, where players' training environs will not be disrupted by major financial hardship or technical barriers. But these grandiose gestures on the part of the developer companies toward the idea of esports as an 'ultimate meritocracy' instead occlude the question of how financially and technically accessible games are in different geographic contexts, and it is necessary to understand such accessibility in order to flesh out the landscape of biopower and biopolitical production in esports.

The idea that esports and video gaming are a meritocracy – an idea that attempts to ignore the primacy of physicality – has been perpetuated with fervour and enthusiasm by some of esports' prominent proponents. Craig Levine, CEO of Electronic Sports League North America (one of the world's largest esports companies, based in Germany), proudly touted in a 2017 promotional article that 'esports has an easy entry point; unlike traditional sports, kids can be any size, shape, gender, and race. It's the ultimate meritocracy' (Getzler, 2017). Esports reporters routinely characterise esports as fundamentally meritocratic, even as they acknowledge that there are glaring flaws and holes in the system to the point where the designation of 'meritocracy' may no longer even be accurate (Van Allen and Myers, 2017; Rose, 2019). The general perception is that video games are a place where structural barriers and disadvantages vanish, and what remains on the battleground is pure, unadulterated skill, be it lightning-speed reflexes, superior aim, or strategic game sense. The fantasy is easy enough to debunk – Christopher A. Paul's book, literally titled *The Toxic Meritocracy of Video Games: Why Gaming Culture is the Worst* (2018), examines gaming culture's adherence to the myth of meritocracy, pointing out how fundamental principles of video game design reinforce that culture: 'Most competitive multiplayer games are designed to enable the best players to win, on the assumption that a video game should be an assessment and adjudication of a player's skill, which is a core tenet of eSports and tournament play' (Paul, 2018: 11). In an industry and fanbase that is still largely dominated by Caucasian, cisgender men, the question of how financial, social, and cultural factors may stand in the way of individual achievement is often ignored. By highlighting South Korean gaming spaces, however, we can see through the veil of meritocracy in a particularly interesting way.

PC *bangs*, which are equipped and specialised for PC gaming, are well known as a staple in the entertainment landscape in South Korea (for a thorough overview of PC *bangs* and their presence in South Korean gaming culture, see Huhh, 2009). They are typically wired for LAN gaming, which means visitors can play with a group of friends on the same network, rather than connecting over the Internet, though they also feature high Internet speeds for online gaming. According to a Korea Creative Content Agency (KOCCA) statistic, there were over 10,600 PC *bangs* in South Korea in 2017 (KOCCA, 2017), which is half the number of the 'golden age' of PC *bangs* in the early 2000s (where the numbers reached over 20,000 PC *bangs*) (Kwon, 2001) but still large enough to assume that most urban neighbourhoods feature a PC *bang* or two.

The importance of PC *bangs* has been emphasised as creating a certain 'game culture' in South Korea, with much of the narrative around

South Korean esports attributing its success to this culture where professional competitive video games are viewed more positively as a career. A lot of the mainstream reporting on South Korean video game culture highlights the significant fanbase that esports players enjoy, the professionalisation of the work, and the competitive, 'cutthroat attitude' (Mozur, 2014). Yet few people appear to acknowledge the most basic physical and material advantage that comes with having a local PC *bang* is accessibility. At its core, a PC *bang* functions much in the same way as a local arcade, in that it features machines equipped for gaming with various pre-installed popular games. A 'budget' gaming PC build, based on North American pricing, can cost roughly US$750 or more (Moore and Walton, 2019), not including an adequate monitor, keyboard, mouse, headphones, and desk, which all together can roughly cost US$1,500–US$2,000. The costs of such a gaming PC set-up, a PC game, and monthly fees of a stable Internet service provider are all entry barriers to PC gaming outside South Korea. in South Korea, a local PC *bang* provides all of the above (sometimes including the game) for the price of roughly US$1 an hour (or cheaper, if you're a member at the PC *bang* and you're spending more time there). At its core, the growth of an esport depends on the size of its engaged player base, and the conditions of gaming access in South Korea mean that the potential for someone to be converted into a member of that engaged player base is much higher.

I highlight PC *bangs* for two main reasons: because I find that they most quickly puncture any over-inflated myths about esports as a meritocratic system, and because they are not brought up often enough in esports conversations about esports growth and longevity. I do not argue that PC *bangs* are the sole factor that differentiates Korean esports training from those of other regions; indeed, it is possible that the militaristic, rigid, age hierarchy that can often be found in Korean male homo-social groups is a significant cultural factor in Korean esports training. However, from the beginning, PC gaming is fraught with economic and material entry barriers – which may perhaps help explain the relative lack of diversity in the North American PC gaming scene in comparison to the console gaming scene (Peterson, 2018) – from the cost of equipment to the presence (or absence) of physical servers for games on entire continents (Grayson, 2017). PC *bangs* go on to serve as important training spaces for fledgeling amateur esports teams looking to go pro; indeed, in 2018, Blizzard Entertainment posted a public notice encouraging PC *bang* owners to support amateur teams looking to enter Open Division (the Tier 3 competition, below OWL and Contenders) for pro *Overwatch* esports by providing them with a physical practice space in the PC *bang*, with Blizzard promising to pay for up to 8,000 hours' worth of PC *bang* fees should the team win (Blizzard PC Bang, 2018). PC *bangs* can and do serve as physical practice spaces where the players can meet in person and practice together, instead of in a 'gaming house' that is now seen as essential for any top tier professional esports team (*ESL Gaming*, 2014). In North America and other places around the world, the journey from aspiring amateur (playing from home) to professional player (living in a gaming house or with a training facility) is still long and arduous, with few physical spaces like PC *bangs* to fill the gap.

The catch, for South Korean esports players, is that despite this infrastructure, their own market is never big enough to hold all of them, and there are plenty of opportunities abroad that provide larger paychecks than domestic teams (Kim, 2016). Xander Torres, an esports writer for *ESPN*, compares South Korea's esports scene to the role that Latin American countries play for Major League Baseball, in that the region acts as a steady supplier of young, fresh talent for teams abroad (personal communication, 17 March 2018, social media direct messages, USA). Emily Rand, another long-time esports journalist currently at *ESPN*, notes that the Korean *League of Legends* scene is a significant feeder specifically for the Chinese esports teams, many of which have significantly larger financial resources at hand than Korean teams. 'Chinese teams scout players directly from the ladder (like Invictus Gaming's Kang 'TheShy' Seung-lok), or from larger teams (like Invictus Gaming's Song 'Rookie' Eui-jin who came from KT Rolster), or minor teams (like JD Gaming's Sung 'Flawless' Yeon-jun)' (personal communication, 12 May 2019, email, USA). She acknowledges that this has had a slow, but inevitably corrosive, effect on Korea's own *League of Legends* esports: 'It's been a slow siphoning off of talent that is now really starting to show. Paired with recent [shifts in the game], it's one of the reasons why South Korea has looked worse last year and this season [at international competitions].'

Despite how much Riot and Blizzard tout their leagues as global, and explicitly model their esports systems on pre-existing global sports competitions, the fact remains that one thing they cannot do is bridge the infrastructural gap of PC gaming that exists across the world. The companies' control over the vertical value chain of esports appears almost absolute and unshakable, yet the truly global league where various regions are adequately represented is far beyond reach – their biopower cannot make up for the difference in accessibility to online PC gaming, which lies at the core of biopolitical production for Korean esports players. Dyer-Witheford and de Peuter refer back to autonomist theorists' works in arguing that 'biopower always rises from the bottom up', and therefore 'there is the possibility of friction between biopower wielded from above and "biopolitical production" rising from below' (Dyer-Witheford and de Peuter, 2009: 223). If we take 'productivity' in the context of esports to mean the honing of skill and the development of team cohesion – the core competitive elements an esports team would need to win matches and attain success – it is the PC *bang* that provides the first, stable stepping stone for casual players to improve their 'productivity' and become more serious competitors. And crucially, this stepping stone is a geographically specific infrastructure, creating a difference that game developer companies cannot easily make up for.

Thus, I highlight PC *bangs* as the physical locus of disruptive biopolitical productivity, a productivity that punctures the meritocracy myth that lies at the heart of the new franchising systems being instituted by biopower-wielding corporations such as Blizzard and Riot. Riot and Blizzard's attempts to shape a global league and create a supposedly meritocratic system of competition are still fundamentally crippled by regionally specific conditions of technological access. A geolocated city system, or a regional team system, depends on the concept of 'locality' to varying degrees for financial revenue; the OWL, which currently holds all of its matches in the Blizzard Arena in Burbank, Calif., plans to have each of its teams relocate to their respective cities by 2020 in order to fully realise its 'esports as local sports' vision. But the stark divide between the conceptual/theoretical and the actual in *Overwatch* and *League of Legends'* esports scenes speak to the still-remaining limits to the companies' 'biopower' in esports.

In applying the framework of biopower to current ongoing changes in PC esports, and specifically to the at-times uneasy relationship that esports leagues have with Korean players, we can see how game developers are attempting to impose a kind of governmentality to the esports landscape in the form of the tightly controlled traditional sports franchising model, as part of their process of vertical integration. We can also see how the pre-existing Korean PC gaming infrastructure provides distinct advantages for Korean players in entering the professional scene, and their outsize presence in the esports landscape (considering the relative size of South Korea compared to other countries with esports scenes) directly interferes with the vision of global esports as envisioned by these developer companies. For the OWL or *League of Legends* to become a global sports league on the scale of soccer or rugby, ideally it would need a more even global distribution of pro players, but – as the London Spitfire, Invictus

Gaming, and Team Liquid can attest – that is simply not the case. While I would not label Korean PC *bangs* as a locus of 'resistance' in any way, the inherent skew in technological infrastructure disrupts the wholly integrated system of value creation that developer companies are attempting to establish, and without significant steps forward in closing that technological gap, it looks likely that Korean players are here to stay in esports.

Notes

[1] Region-locking is a rule found across multiple esports titles that either (a) requires players to have citizenship or at least residency in the region in which they compete, or (b) limits the number of 'non-local' players that an esports team can have. 'Region' in this context can refer to a continent or a country, depending on how the esports league itself chooses to separate the different geographic locations in which pros compete, and the qualifications for 'residency' here are not identical to legal residency. The major regions for *League of Legends* esports, for instance, are North America, Europe, China, and South Korea; Korean players who seek to play in North America or China must play in that region for three years in order to claim 'resident' status. Region-locking has both proponents and opponents in esports, with proponents saying that it prevents true competition from happening, whereas opponents argue that region-locking is the best way to foster regional talent and a strong local fanbase. Region-locking is usually instituted in order to guarantee some degree of regional diversity at world tournaments; in *League of Legends*, for example, it was first instituted in 2014 after LMQ, a team originally founded to compete in China, moved wholesale to North America for the easier competitive landscape and eventually won a spot to represent North America at the World Championship (Rhea, 2014). It was instituted for similar reasons in *StarCraft II* esports after Korean players moved to North America to represent the region, as the Korean competitive scene for *StarCraft II* proved much too cutthroat and demanding (Yabumoto, 2017). For a roundtable discussion on the pros and cons of region-locking in *League of Legends*, see Li (2017).

[2] Inasmuch as it has created (annual) player salary minimums of US$50,000, the OWL has been quite successful in establishing a degree of financial security for its players, but the long-term sustainability of these salaries has come into question. Moreover Blizzard has also come under criticism for its treatment of its second- and third-tier tournaments, which are direct feeder competitions for the first-tier OWL. Essentially, financial security has been achieved for the top-tier players signed to an OWL team, but for the large number of players in lower-tier *Overwatch* competitions (Contenders, Open Division), there is very little financial security. For an in-depth discussion of this issue, see Partin (2019).

[3] 'Third party' here refers to any entity that is not the game developer itself.

[4] Admittedly, a complicating wrinkle in this US-centric landscape is the fact that the Chinese corporation Tencent owns the majority of shares in Riot Games, 5 per cent stakes in Blizzard and Activision, as well as in numerous other game companies.

[5] As noted by the authors, the phenomenon has counterparts in other games such as *Everquest*, *Ultima Online*, *Dark Age of Camelot*, and so on.

[6] The represented cities are: Atlanta, Boston, Chengdu, Dallas, Guangzhou, Hangzhou, Houston, Los Angeles (with two teams), New York, Paris, San Francisco, Seoul, Shanghai, Toronto, and Vancouver.

[7] The five 'premiere' professional leagues for *League of Legends* are: North America, Europe, South Korea, China, and Taiwan–Hong Kong–Macau. In addition, there are leagues for Vietnam, the Commonwealth of Independent States, Turkey, Brazil, Latin America, Japan, Oceania, and Southeast Asia.

[8] I ask these specific questions for a couple of different reasons. Firstly, as I discuss in this chapter, because of the gap between what the esports leagues themselves appear to envision – global, meritocratic competition – and the reality on the ground, in which Koreans appear to be 'over-represented' considering their population size and available capital. Secondly, the questions are not simply derived from my own observations, but are also critical topics that some of the esports communities themselves have raised frequently throughout recent years, leading to fraught debates about how North American, European, and Oceanic teams should foster more 'domestic' talent instead of 'importing' Korean players, and what the esports leagues themselves can do to prevent such a skewed professional player base. While some of these online debates also feature significant amounts of xenophobia and racism, with one prominent esports commentator at one point referring to Korean esports players as 'termites' spreading in the esports scene, they also reflect the significant anxieties that various esports regions have in terms of fostering and cultivating a healthy, growing esports ecosystem.

References

Blizzard PC Bang (2018) 'Bŭllijadŭ esŏ kongsik chuch'oe hanŭn opŭn dibijŏn PC bang kudanju rŭl mojip hapnida', [Blizzard seeking PC Bang owners for official open division tournament], 2 January, Available from: https://pcbang.blizzard.com/ko/notice/21353260 [Accessed 5 January 2021].

Burk, D. (2013) 'Owning e-sports: Proprietary rights in professional computer gaming', *University of Pennsylvania Law Review*, 161: 1535–77.

Carpenter, N. (2018) 'South Korea's prestigious OGN Overwatch Apex tournament is no more', *Dotesports*, 5 January, Available from: https://dotesports.com/overwatch/news/overwatch-apex-ogn-blizzard-19944 [Accessed 5 January 2021].

Cuevo, C. (2019) 'Immortal fighting games: How one of the oldest esports genres has survived generations', *Inven Global*, 8 June, Available from: https://www.invenglobal.com/articles/8357/immortal-fighting-games-how-one-of-the-oldest-esports-genres-has-survived-generations [Accessed 5 January 2021].

D'Anastasio, C. (2019) 'Shady numbers and bad business: Inside the esports bubble', *Kotaku*, 23 May, Available from: https://kotaku.com/as-esports-grows-experts-fear-its-a-bubble-ready-to-po-1834982843 [Accessed 5 January 2021].

Dyer-Witheford, N. and de Peuter, G. (2009) *Games of Empire: Global Capitalism and Video Games*, Minneapolis: University of Minnesota Press.

Ellis, S. (2016) 'Esports is growing up: IP law and broadcasting rights', *ESPN*, 25 January, Available from: https://www.espn.com/esports/story/_/id/14644531/ip-law-broadcasting-rights-esports [Accessed 5 January 2021].

ESL Gaming (2014) 'Team houses and why they matter', 6 January, Available from: https://www.eslgaming.com/article/team-houses-and-why-they-matter-1676 [Accessed 5 January 2021].

Fogel, S. (2018) 'ESPN, Disney, ABC airing overwatch league', *Variety*, 11 July, Available from: https://variety.com/2018/gaming/news/espn-disney-abc-overwatch-league-1202870413 [Accessed 5 January 2021].

Getzler, W. (2017) 'eSports and kids: Welcome to the big league', *Kidscreen*, 13 September, Available from: http://kidscreen.com/2017/09/13/welcome-to-the-big-league/ [Accessed 9 February 2021].

Grayson, N. (2017) 'It isn't easy being an overwatch fan in Africa', *Kotaku*, 13 September, Available from: https://kotaku.com/it-isn-t-easy-being-an-overwatch-fan-in-africa-1806133900 [Accessed 5 January 2021].

Huhh, J-S. (2009) 'The "bang" where Korean online gaming began: The culture and business of the PC *bang* in Korea', in L. Hjorth and D. Chan (eds) *Gaming Cultures and Place in Asia-Pacific*, New York: Routledge, pp 102–16.

Isaakov, E. (2016) 'OGN controversy breakdown', *Dotesports*, 15 January, Available from: https://dotesports.com/league-of-legends/news/ogn-controversy-breakdown-11473 [Accessed 5 January 2021].

Jin, D.Y. (2010) *Korea's Online Gaming Empire*, Cambridge, MA: The MIT Press.

Kim, A. (2016) 'Report: LCK players demanding higher pay because of inflated market', *Slingshot*, 5 December, Available from: http://slingshotesports.com/2016/12/05/report-lck-players-demanding-higher-pay-because-of-inflated-market/ [Accessed 1 July 2019].

KOCCA (Korea Creative Content Agency) (2017) '2017 White paper on Korean games', Seoul: Korea Creative Content Agency.

Kwon, M. (2001) 'PC bang 2 man'gae yukbak' [PC *bangs* reach 20,000 in number], *Maeil Business Newspaper*, 12 July, Available from: https://www.mk.co.kr/news/home/view/2001/07/177350/ [Accessed 5 January 2021].

League of Legends (2018) '2018 events by the numbers', Available from: https://nexus.leagueoflegends.com/en-us/2018/12/2018-events-by-the-numbers/ [Accessed 5 January 2021].

Lee, H. (2019) 'Sŏ'uldae ŭidae iphakpoda puro keimŏ toegi ŏryŏwŏ' [It's harder to become a professional gamer than it is to get into Seoul National University medical school], *Chosun Ilbo*, 29 June, Available from: http://news.chosun.com/site/data/html_dir/2019/06/28/2019062801980.html?utm_source=naver&utm_medium=original&utm_campaign=news [Accessed 5 January 2021].

Li, X. (2017) 'Should Riot re-think region locking? Immortals' CEO Noah Whinston doesn't think the current rules go far enough', *Dotesports*, 26 July, Available from: https://dotesports.com/league-of-legends/news/riot-league-region-locking-15819 [Accessed 5 January 2021].

LoLesports Staff (2016) 'Changes to the Interregional Movement Policy (IMP)', lolesports.com, 2 August, Available from: https://nexus.leagueoflegends.com/en-us/2016/08/changes-to-the-interregional-movement-policy/ [Accessed 1 July 2019].

McCrea, C. (2009) 'Watching StarCraft, strategy, and South Korea', in L. Hjorth and D. Chan (eds) *Gaming Cultures and Place in Asia-Pacific*, New York: Routledge, pp 179–93.

Moore, B. and Walton, J. (2019) 'Budget gaming PC build: The best parts for an affordable gaming PC in 2019', *PC Gamer*, 8 April, Available from: https://www.pcgamer.com/pc-build-guide-budget-gaming-pc/ [Accessed 5 January 2021].

Mozur, P. (2014) 'For South Korea, e-sports is national pastime', *The New York Times*, 19 October, Available from: https://www.nytimes.com/2014/10/20/technology/league-of-legends-south-korea-epicenter-esports.html [Accessed 5 January 2021].

Pannekeet, J. (2019) 'Global esports economy will top $1 billion for the first time in 2019', *Newzoo*, 12 February, Available from: https://newzoo.com/insights/articles/newzoo-global-esports-economy-will-top-1-billion-for-the-first-time-in-2019/ [Accessed 5 January 2021].

Partin, W. (2018) '"StarCraft II": How blizzard brought the king of esports back from the dead', *Variety*, 13 July, Available from: https://variety.com/2018/gaming/features/starcraft-ii-esports-history-1202873246/ [Accessed 5 January 2021].

Partin, W. (2019) 'The esports pipeline problem', *Polygon*, 11 July, Available from: https://www.polygon.com/features/2019/7/11/18632716/esports-amateur-pro-players-teams-talent-process [Accessed 5 January 2021].

Paul, C.A. (2018) *The Toxic Meritocracy of Video Games: Why Gaming Culture is the Worst*, Minneapolis: University of Minnesota Press.

Peterson, L. (2018) 'Why aren't more black kids going pro in esports?', *The Undefeated*, 27 March, Available from: https://theundefeated.com/features/why-arent-more-black-kids-going-pro-in-esports [Accessed 5 January 2021].

rainha (2015) 'From Sejong the great to SKT T1 – How Korean culture sets the region apart', *Dotesports*, 26 October, Available from: https://dotesports.com/league-of-legends/news/from-sejong-the-great-to-skt-t1-how-korean-culture-sets-the-region-apart-8117 [Accessed 5 January 2021].

Rand, E. (2018) 'Culture shock: The multinational mosaic of Overwatch League', ESPN, 23 July, Available from: http://www.espn.com/esports/story/_/id/24168436/culture-shock-multinational-mosaic-overwatch-league [Accessed 5 January 2021].

Rhea, M. (2014) 'Riot announces new rules about regional movement', *RedBull*, 6 September, Available from: https://www.redbull.com/us-en/riot-announces-new-rules-about-regional-movement [Accessed 5 January 2021].

Rodriguez, V. (2019) 'Dota 2 international prize pool: Money over the years, *DBLTAP*', 1 February, Available from: https://www.dbltap.com/posts/6286261-dota-2-international-prize-pool-money-over-the-years [Accessed 5 January 2021].

Rose, V. (2019) 'Esports' biggest trends and how they're going to shape 2019', *GameDaily.biz*, 8 January, Available from: https://gamedaily.biz/article/493/esports-biggest-trends-and-how-theyre-going-to-shape-2019 [Accessed 5 January 2021].

Taylor, T.L. (2012) *Raising the Stakes: E-sports and the Professionalization of Computer Gaming*, Cambridge, MA: The MIT Press.

Team Liquid (2019) 'Partners', *teamliquid.com*, 24 July, Available from: https://www.teamliquid.com/partners [Accessed 5 January 2021].

Van Allen, E. and Myers, M. (2017) 'The state of esports in 2017', *Kotaku*, 14 December, Available from: https://compete.kotaku.com/the-state-of-esports-in-2017-1821304905 [Accessed 5 January 2021].

The Versed (2017) 'How do eSports teams make money?', 28 March, Available from: https://www.theversed.com/3554/how-do-esports-teams-make-money/ [Accessed 5 January 2021].

Volk, P. (2016) 'Riot's residency rules are changing: Here's how and what it means', *Rift Herald*, 1 August, Available from: https://www.riftherald.com/competitive/2016/8/1/11695806/lcs-residency-rule-requirement-change-homegrown-explainer [Accessed 5 January 2021].

Webster, A. (2018) 'Why competitive gaming is starting to look a lot like professional sports', *The Verge*, 27 July, Available from: https://www.theverge.com/2018/7/27/17616532/overwatch-league-of-legends-nba-nfl-esports [Accessed 5 January 2021].

Witkowski, E. (2012) 'On the digital playing field: How we "do sport" with networked computer games', *Games and Culture*, 7(5): 349–74.

Wolf, J. (2017) 'NRG signs 17-year-old Overwatch pro sinatraa for $150K', *ESPN*, 7 September, Available from: https://www.espn.com/esports/story/_/id/20564135/nrg-signs-17-year-old-overwatch-pro-sinatraa-150k [Accessed 5 January 2021].

Yabumoto, J. (2017) 'The good and bad of StarCraft II's esports region lock', *Gamecrate*, 29 September, Available from: https://www.gamecrate.com/good-and-bad-StarCraft-iis-e-sports-region-lock/17226 [Accessed 5 January 2021].

Zacny, R. (2016) 'The complicated, messy future of esports', *Kotaku*, 21 January, Available from: https://kotaku.com/the-complicated-messy-future-of-esports-1754370671 [Accessed 5 January 2021].

8

South Korea's Esports Industry in Northeast Asia: History, Ecosystem and Digital Labour

Peichi Chung

Introduction

This chapter studies the formation of the esports gaming network in Northeast Asia by examining the case of South Korea in order to explore the role that its government plays in developing regional esports gaming culture. Since the 2000s, South Korea's game industry has sparked a new wave of content circulation for transnational games developing in East Asia. Thanks to the advancement of networked technology, South Korea also now leads in setting global trends in esports professionalisation (Taylor, 2012). A view of early esports history shows that in the late 1990s, the popular household game *Starcraft* kickstarted the country's gaming revolution. Indeed, many amateur *Starcraft* players made the leap into professional gaming to become full-time pro players. In 2000, the South Korean government established its flagship esports office, the Korean Esports Association (KESPA). The next year, 131 professional players registered to become members of online game teams (Jin, 2010). In addition to policy factors, 'corporate incentives also led to the popularity of esports in South Korea' (Taylor, 2012: 25). Since 2000, Samsung Electronic has internationalised esports by hosting annual worldwide 'Olympics' like tournaments, called the World of Cyber Games (WCG). Other technology conglomerates such as SK Telecom, Korean Telecom, Wemade and Jin Air nurtured the domestic esports industry by sponsoring teams in professional leagues. By the 2010s, attending events in esports arenas were becoming a part of an esports fan lifestyle in cities like Seoul. Finally, by 2014, the *New York Times* went so far as to describe esports as the national pastime of South Koreans (Mozur, 2014). Simply put, esports has grown rapidly in the country's economy and culture.

As a consequence, today 'pro gamers are considered to be important components of Korea's digital economy and culture-driven Korean society' (Jin, 2010: 82). This chapter offers a case analysis that shifts away from a corporation-centred approach to an alternative method that reveals the country's player-driven esports policy model of innovation. In South Korea, young gamers can gain easy access to fast-speed computers through PC *bangs* (Huhh, 2009). The country's esports fans hone their gaming skills by spending long hours playing with friends. Today, South Korea is recognised as Asia's esports hub. However, despite its leadership in esports development, controversy among Koreans exists because the esports community continues to struggle with challenges from the Korean public and institutions worldwide regarding the negative impact of game addiction. In 2020, the global esports industry continued its rapid expansion, reaching US$1.1 billion in revenue. However, the esports community in Korea still has to battle the stereotype of teenage students playing these games too much at the expense of their education. In addition, external conflict also occurred when in September 2018, the World Health Organization (WHO) officially included gaming disorder as a disease in its International Classification of Diseases (WHO, 2018). In December 2019, the International Olympics Committee (IOC) released its disapproval of a proposal to include esports in the upcoming Olympic Games (see also Introduction). The IOC declared that esports would not be considered an Olympic sport due to its connections to mental health problems and in-game violence (International Olympic Committee, 2019).

The findings in this chapter are based upon three years (2017–19) of research fieldwork that I conducted in the South Korean cities of Seoul, Gwangju, Naju, and Busan. More than twenty in-depth interviews were conducted during this fieldwork period with enthusiasts who are passionate about esports. They include esports professionals who work for esports companies and esports associations, as shoutcasters, as game channel streamers, as coaches, and, of course, as amateur and professional players. The boundary between amateur and professional is blurred, as the shoutcaster interviewed, for instance, is a social media esports competition announcer who often hosts in Korea's game TV, Ongamenet, and owns channels on YouTube and Twitch. The analysis of these interviews contextualised their voices within the larger societal structure to highlight the dilemmas that esports professionals sometimes face in the development of their career paths. Thus, a labour-centric approach is needed to understand the efforts required to pursue esports professionally. This approach will positively contribute to ongoing debates about issues of public welfare in esports governance, both in South Korea and beyond.

Background of esports in Northeast Asia

South Korea's uniqueness in esports development can be understood within a larger framework when compared to Japan and North Korea. Comparably, South Korea is recognised as the forerunner in esports development. Japan is another strong gaming nation in Northeast Asia. Comparatively, in 2018, South Korea possessed 28.9 million gamers. Meanwhile, Japan has the largest gaming population, with 68.7 million players.

South Korea, as expected, possesses the highest number of esports teams in the region: SK Telecom T1, MVP Phoenix, Samsung Galaxy, SK Telecom T1 K, MVP Black, ROX Tigers, KT Rolster, GenG, Ballistix, and Tempest are the top-earning esports professional teams registered in the country. As of March 2020, South Korea has a total of 217 esports teams, among which SK Telecom T1 ranks the highest in terms of earnings – US$5,622,438 (eSportsflag, 2020). Most South Korean esports teams play games that are developed by companies based in California. Competing titles include *Starcraft* (1998), *League of Legends* (2009), *Overwatch* (2015), *Clash Royale* (2016), *Honor of Kings* (2016), and *Player Unknown's Battlegrounds* (2016). Table 1 shows that professional players in South Korea outnumber professional players in Japan by a factor of sixteen. It is noted that the highest earning players in each country, Faker and Feg, earn comparable salaries. However, further comparisons reveal that top-performing South Korean esports players generally earn larger salaries than their Japanese counterparts.

In Japan, an organised esports network was not launched until 2018, when a coalition of game publishers and other esports companies established the Japanese Esports Union (JeSU) (Hoppe, 2019). JeSU combines three independent esports institutions: the Japanese Esports Association, the Esports Promotion Organization and the

Table 1: Esports teams, players and earnings in South Korea and Japan in March 2020

By country	Number of esports teams	Number of pro players	Most played game	Top three players in terms of earnings	Total earnings (US$)
South Korea	217	115	*Starcraft*	Faker	1,255,464
				Duke	954,620
				Wolf	913,084
Japan	67	1,890	*Overwatch*	Feg	1,003,000
				Tokido	486,404
				Daigo	238,984

Source: https://www.esportsearnings.com and https://esportsflag.com

Japan Esports Federation. Japan's pro-gaming environment includes a strong Japanese corporate presence. Corporate dominance has prevented many talented Japanese gamers from embracing game titles produced by game companies like Riot Games and Blizzard that are based in California. It has been noted that reputable companies such as Nintendo, Sony, Capcom, Namco Bandai, and Konami own many successful titles such as *Mario* (1981), *Street Fighter* (1987), and *Pokémon* (1996). Being the third-largest gaming nation in the world makes Japan a difficult esports market to penetrate. JeSU issues licenses to hold pro gamer tournaments with prize money (Ashcraft, 2018). Most of the top prize awards have gone to Japanese titles, like *Street Fighter* (1987) and *Shadowverse* (2016). Many attribute Japan's late start in cultivating its national esports environment to the state's strict laws governing the businesses of gaming and entertainment. For example, an anti-gambling law 'Act Against Unjustifiable Premiums and Misleading Representations' prevents gamers from winning monetary prizes from corporations. Only when the Japanese government sees esports grow large enough to project a form of soft power and to showcase Japan in Olympic competition will the government adjust its regulatory approach. With the current government regulations, it will be hard to speed up the country's esports development.

Extended background information about gaming in North Korea shows that the country's cyber-force trains hackers to sell illegal software and steal personal data from the South Korean gaming network (Kim, 2011). The state sponsored the development of a few phone games to keep Internet users from watching smuggled American and South Korean media programmes from China. The stories in most of these games are based on propaganda that complies with the demands of political power in Pyongyang. The antagonists in these games are enemies in South Korea, Japan, or the United States. In particular, the game, *Boy General*, tells the story of a young general who tries to save his country (a representation of North Korea today) from a villainous alliance between China's Tang Dynasty (7th–10th centuries) and a southern Korean Kingdom (CNBC, 2015). Other phone games include *Tank Battle*, *Bubble Popping*, and *Baby Piano*. *Pyongyang Racer* is a racing game that showcases North Korean tourism by driving the player through the streets of Pyongyang.[1] The game incorporates a unique music style and can be played at the web site of North Korea's tourist company, Koryo Tours. Games made by the Japanese Sega Corporation and arcade games also circulate in North Korea. Despite this gaming presence, gamer skills in North Korea are considered less advanced than in countries like Japan and South Korea.

The literature on labour between play and work

The literature of game studies discusses different stages of gaming between play and work. Related questions include: Is gaming an authentic form of play? Does it create value? In the digital capitalist system, when will playing games for leisure become a form of work? Why are we seeing more and more gamers who desire to play games competitively as professional players? Does this phenomenon reflect a new form of exploitation that embodies false consciousness and alienation with regards to the labour conditions in esports?

Lund identifies play and work as distinct arenas of gaming activity. He writes: 'playing is engaged in form of the pleasure of the activity in itself and work is for the satisfaction of a need by the production of certain use value' (2014: 763). For Kucklich (2005), play and work are interlinked, and therefore have led to the emergence of the new hybrid form of labour known as *mudding*. Indeed, Kucklich (2005) goes further, declaring that 'the precarious status of mudding' defines it further as 'a form of unpaid labour'. Mudding is a leisure activity, but it also produces free labour and celebrity players. Kucklich coins the term *play-bour* to specify a new powerful player status for those fans who provide innovative feedback to game developers. This new labour function empowers gamers to alter the old exploitative relationship between corporations and gamers from its traditional Marxist labour context.

Most pro players begin their gaming career by playing for leisure. The estimated career path for professional players starts at the age of 15 and ends at around 25. Various incentives might be involved in play, be they boredom, fun, or socialisation. Gaming experiences include factors such as immersion and identity. People may play games in order to discover more about their stories or to discover the various components that reflect the aesthetics in game design. A good game inspires players' subjective feelings to create the fantasy of a better world (Bartle, 2004). As the player negotiates with structured actions to overcome obstacles encountered in the completion of a story (Rodriguez, 2006), esports adds competition and high-power performance to the meaning of play (Witkowski and Manning, 2018). In esports, the idea of 'free play', created by Huizinga (1955), requires 'physical and mental abilities in the use of information and communication technologies' (Ratliff, 2015: 32).

Regarding work, the idea of class exploitation and corporate ideology masked by the form of gaming points to the problem of work inequality in digital space (Sholtz, 2013). Mudders, wiki page writers, esports players, and Instagram celebrities all encounter the similar problem of labour exploitation in generating value (Sholtz, 2013). Employment has changed as a concept because it has become increasingly difficult to distinguish play, consumption, and production from work in today's digital platforms, where the labour–corporation relationship can now involve unpaid productivity for gamers who volunteer to test games to offer design feedback to game companies for free. Work can be glamorous, but most of the time it is precarious (Terranova, 2013). Fuchs (2013) considers this new relationship an ideology of the participatory web. Social media platform companies lure consumers to participate as means to create surplus labour. In this scenario, users become 'useful labour' to a system of capitalist oppression and discrimination. In addition, the racialisation of Chinese gold farmers also addresses the political reality of discrimination in digital labour when one observes who 'works' in the virtual economy of games like *World of Warcraft* (2004) (Nakamura, 2013).

Work in esports is most deeply analysed in the book *Raising the Stakes: E-Sports and the Professionalization of Computer Gaming*. In this book, Taylor (2012) addresses one of the dilemmas that professional players often face in their digital workspace. Many professional players have had to leave esports to seek jobs outside their profession. According to Taylor, professional players are power gamers who have developed skills and tactics, and accrued a deep knowledge of the gaming system. Players possess expertise in how to interact with data and machines, as well as how to handle social interactions while competing in digital environments. The capabilities of professional gamers develop from long hours of practice playing team games in the long process by which their gaming transforms from play to work. Taylor notes that once esports becomes an occupation, pro players are ready to compete like professional athletes. As they win consideration for top-level player positions, their lives assume work schedules complete with team players, coaches, and sponsors. And as they travel to tournament competitions and attract fan support that brings glamour to their profession, professional players also sometimes must struggle with more difficult realities such as dealing with the 'tricky domestic situation with family and friends who may not entirely understand what they are trying to do' (2012: 132).

This 'labour' in turn highlights features of the transformative process between play and work in esports. In the article 'Digital labour studies go global', Casilli (2017) argues for a framework that makes 'work visible'. The subjective experience of the power gamer expresses a mythic aspect with regard to the play-work transformation in esports professionalisation. Only a few lucky esports enthusiasts are ever able to achieve top-level player careers (Taylor, 2012). When this transformation does occur, it involves to a large degree labour conditions that pro players have imposed on themselves. Part of the reward for enduring these conditions lies in the desired achievement of some expressed goal, such as winning a particular tournament or competition. Professional players gain advanced skill development and learn to master strategy by interacting with AI-based computer systems (Canning and Betrus, 2017). This advancement is part of a reskilling and upskilling process that is unique to data science and esports. A study of player roles in multi-player game, *Dota 2*, for example, notes a deep-learning outcome as *Dota* players perform team-based behaviours and movements in game matches (Eggert et al, 2015). This demonstration of skill enhancement sheds light on the further need to examine esports and its role in facilitating workplace interaction and management in the world's unfolding data society.

The South Korean esports industry

History of KESPA

A review of the history of esports development reveals how two public organisations built South Korea's esports ecosystem by developing nationalising and globalising infrastructures. Specifically, the effort to create a national esports culture can be traced back to the domestic and regional governance strategies designed by KESPA.

Since its establishment, KESPA has been noted for its development of a pro-gaming licensing system and partnership programmes that work with various industry stakeholders. This includes the state: The KESPA office is an extended arm of the Ministry of Culture, Sport and Tourism. Altogether, they promote esports as a state-sanctioned healthy sport. In 2006, the Ministry of Culture, Sport and Tourism enacted the Act on the Promotion of Esports. This policy initiates the country's long-term government plan to invest in the improvement of infrastructure and award subsidies. Recent KESPA promotions encourage family esports events, seeking to inspire family participation in public esports events in hopes of changing the public perception of esports away from the perception of a lone player in isolation to the alternative of a sport that helps build social networks. Related esports regulations such as the Shutdown Law – which was implemented in 2011 and supported by both the Ministry of Culture, Sport and Tourism and Ministry of Gender Equality and Family – aims to curb Internet addiction among under-aged players. In 2017, the government passed another esports law that bans paid boosting services under the Game Industry Promotion Act. In competitive gaming, boosting is an unethical practice that allows players to use various extended supports to inflate the performance of their game play. Boosters who commit this crime are subject to fines of up to US$18,000 and two-year suspended sentences.

The achievements of the KESPA include two decades of continuous efforts in business extension that promote and connect professional players to various industry stakeholders in Korean society. KESPA has worked with professional team sponsors to nationalise *Starcraft* and *League of Legends* competitions. The office collaborates with game developers at Riot Games, Nexon, and Valve Corporation to host game events for different gamers. Among all the tournaments, the KESPA Cup perennially attracts professional players from not only South Korea, but also from abroad. Like other traditional sports, KESPA also collaborates with provincial governments to develop amateur talent. As industry stakeholders tend to see esports as opportunities to expand business and extend brand recognition, KESPA carves out a career path of industry development for professional players. It has control over the qualifications of a professional player through its award of official KESPA approval, and can decide which players can participate in matches. It also has the power to ban players if rules are not followed during competitions. KESPA's influence extends to broadcasting; the organisation was once involved in a dispute with Blizzard and broadcasters OGN and MBC Game over broadcasting ownership right in the late 2000s. However, ever since KESPA was ruled out to own management right on tournament broadcasting content after agreement with game developers, it has further focused on nationalising South Korean esports within the regional context of the Asian Games. The office has signed MOUs with its counterparts in China, Vietnam, Taiwan, Japan, North America, Sweden, and Mexico.

The regional strategy of KESPA can be seen in its promotion of esports in the PC game genre. An interview with an experienced KESPA policy officer, Mr A,[2] shows that KESPA understands the strength of South Korean esports players in the PC game genre, while the US, Japan, and Europe excel at console games. To him, KESPA focuses on developing an Asian strategy by actively promoting esports as an official sport in the Asian Games. Mr A describes the organisation's effort in this direction since 2009, when KESPA initiated a conversation about esports inclusion in the Games. In 2018, esports became a demonstration sport in the 18th Indonesia Palembang Asian Games. The 2019 Southeast Asian Games awarded six esports medals. These games were accredited by the Asian Electronic Sports Federation. As more participating institutions in the Asian region joined the push to make esports a formal sport, Mr A shares his observation of esports expansion from the national to regional scale from the perspective of KESPA in the following interview:

> We [KESPA and the Ministry of Culture Sport and Tourism] need to communicate issues with the Ministry of Education. We used to talk to the sport department. We need help from education side […] In 2018, we just sent in two games, *League of Legends* and *Startcraft*. Because in Korea we are small, we are good at PC genres. We are not strong in mobile and console. But, in Asia, Asian Games needs to take into [consideration] the whole region of Asian game play. We need more titles to give [a] chance to Middle Asia. But formerly, in the 2016 Asian games, I chose six games and selected each game from each genre. We considered an Asian esports solution. Console genre is strong on the west side of Asia and the PC genre is strong on the east side […] Because Korea is very small, I need Asia's support to make the esports story. It took me ten years of experience to explore the way to enter the Asian Games. We dreamed about this from 2009. [Since then] we started talking. Now we can communicate. Korea has the most well organised and well known [esports gaming landscape]. We

are glad that now it expands to other countries, and more industry resources from other countries have joined. (Mr A, interview with the author, July 2018)

History of IESF

The International Esports Federation (IESF) is a non-profit esports organisation that started in close collaboration with KESPA. The organisation formed an alliance with eight other member nations to promote esports as a legitimate sport within the global community. The member countries are South Korea, Taiwan, Vietnam, Denmark, Germany, Austria, Belgium, the Netherlands, and Switzerland. From 2009 to 2019, the IESF hosted annual championship tournaments around the world, in countries such as South Korea, Sweden, Taiwan, Serbia, and Japan. Within South Korea, the strongest support appears to exist in Busan, so much so that the IESF moved its office there from Seoul in response to the city government's initiatives demonstrating support. The IESF has developed a well-connected international network of stakeholders including SK Telecom, World Cyber Arena, Ali Sports, and Riot Games. In 2012, the federation held its first women's tournament. In 2020, it extended its regional network to include the Asian Esports Federation, and its global reach had extended to include sixty nations on six continents.

A process of transformation from play to work is seen in the organisational growth of the IESF. The IESF seeks esports professionalisation by standardising the rules for global esports competition. This agenda serves the purpose of creating a space for players in the global esports industry at a time when players are also seen as under the control of game corporations. The IESF works to promote esports to the legitimate status of a sport worthy of recognition by an international sports committee. This push primarily helps to set up a business model as the IESF expands its collaborating partners in the global esports market. Additionally, it also shows a bottom-up approach to creating a public space that accommodates both amateur and professional players in the global esports governance ecosystem. The concentration on player-driven agendas allows the IESF to distinguish its non-profit position when negotiating with global publishers, broadcasters, event sponsors, and governments. This perspective allows it to focus on extending the potential career cycles of professional players. In 2016, the IESF established the International Esports Academy in order to set up online educational programmes that grant certificates to players who are interested in working as tournament referees. One of my interviewees analyses the IESF's player-focused approach and concludes that it is a reflection of a trend evolved from the changing nature of today's young generation. He observes that fans have fully embraced esports because they show the young generation how to connect socially with each other. Because of this trend towards the rapid growth of esports culture, the IESF sees a need to look for a way to fit esports into the global structure of sport so that players can be like athletes who are protected by initiatives agreed to by the international sports community. The interviewee further explains the organisation's player-based principles:

> We want to set up quite a good brand for esports tournaments that can be compared to the Olympic brand. At the same time, we want to start some kind of intellectual property business that can be a stadium business, an academy, and some good campaigns to undertake global redistribution. The basic concept [for all businesses] is all about the players. It is [that] the player is the key stakeholder who actually makes the world of esports work, not only the professional players, but also the amateur players. Even these amateur players will eventually become professional players, and professionally will retire. This is the ecosystem that should have the player's pool [sic]. Pushing universal values for good campaigns, we can draw on some channels to actually make that happen. People will realise that what we do is really necessary work that should be done in the industry. In the past, it is [sic] the sport industry. Eventually the game companies will realise that this is the area that everyone should be involved in as well. (Mr B, interview with the author, July 2018)

The IESF has therefore developed rules to standardise player qualifications and behaviour. It regulates criteria to define matches, referees, stadiums, uniforms, broadcasting rights, and event sponsors. The organisation has even proposed anti-doping rules as part of its campaign to promote esports as a healthy game. Because of the organisation's pioneering role in developing rules and regulations for competitions, the IESF redefines the different corporation and player relationships in esports.

There is a need to implement further policy measures on welfare development in the global esports market. Such initiatives correspond to laws proposed in some Western countries. The regulation of players

– like in the California Code of Labor, Fair Labor Standard Act (US), National Labor Relations Act (US), and Digital Republic Bill (France) – point to the increasing attention being paid to the protection of professional leagues, as young players handle contracts and relevant issues such as earnings, sponsorship, team organisation, and streaming as a part of their professionalisation development.

Ecosystem

South Korea's player-driven esports innovation is also prevalent in all perspectives of the esports industry, as we see in an expansion of value chains in the making. Most of my interviewees had all dreamed of careers in professional gaming during their youth. While some are still trying to achieve this dream by staying on amateur teams and participating in tournaments hosted by the Korean government and game companies, a substantial part of the interviewees have gone further and continue creating alternative esports career paths after leaving their professional gaming careers. Some of my interviewees became shoutcasters, a new profession that involves shouting commentary over video gaming streams. They operate social media channels to showcase games in the first-person shooting genre. Some of the interviewees are now esports coaches, lead amateur teams, and teach esports courses in universities. The issue of players who leave professional gaming but remain in the esports industry is an essential one to raise, especially given the new labour conditions developing in global esports. How much should this be an issue of concern for game corporations like Riot Games, Tencent and Blizzard? Why do retired players stay in the industry by creating new businesses or waiting for work opportunities in such a new industry?

Gamer-led innovation shows up in five areas of value-chain expansion in my fieldwork. The five areas include professions in the businesses of game publishing, sports, education, media, and regulation. To the present day, most discussions about the esports ecosystem develop within the context of market return, as esports has a major impact on brand promotion for esports games. However, as we examine the growing interests of various parties invested in the esports market, we see that 'monetisation' has to involve a good balance among factors such as game publishing, athletics, education, media, and regulation. For instance, in Northeast Asia, there is a growing interest in curriculum design for esports education. There is also a growing need to look into the regulatory body developed between corporations and government. What are the interests that we should protect if esports is a new type of global sport that only a few corporations now possess the power to control under the global intellectual property rights system?

The story of another interviewee, Mr C, offers a glimpse into the growing esports division within a game company. Because of his passion for esports, Mr C works in the esports division of a large game company in Korea as a team manager. Mr C's job description includes growing the presence of the company's game by launching a local professional esports league and hosting small-scale tournaments in Korea and throughout Asia. Mr C tries to create grassroots esports by working with publishing partners and venue sponsors to increase the game's market share. He is responsible for designing rules that help to define the company's tournament system as a part of the effort to form a global league. However, the success of this value chain remains uncertain, even as game companies continue to invest heavily in developing esports as a new 'cultural business'. He mentions the company's effort to drive 'monetisation' in the following interview:

> I agree that esports is a very important factor in game marketing, including enhancing the brand of the publisher. The consensus among game developers and publishers to promote esports is high. But to a certain point, we are not able to know exactly how much esports contributes to our revenue gain. The revenue increase is not only from the effort of esports promotion. It is also helped by other marketing strategies. Main game companies like Valve invest so much in esports, even though it is hard to measure the impact. But big companies consider esports as a cultural business, not just marketing activities. Since we are on the road to a new cultural business, big game companies are holders of the IP rights. We just continue to invest in this new business. (Mr C, interview with the author, July 2016)

In my fieldwork, I have observed that the challenge in creating a visible line of monetisation in value chain expansion is mostly seen in the management of an esports team. Several of my interviewees work as professional players, team coaches, and owners of a team that compete in the secondary league in South Korea. The hierarchy among player, coach, and sponsor became clear in the course of my interviews with professional gamers. The daily routine of a professional gamer involves an intensive schedule of practice and match play with arranged guest teams. Most of the players are young. Many have quit secondary school or university to join a professional team. Winning a competition remains a constant and foremost goal in the minds of these young professional gamers. Living together almost twenty-four hours a day with teammates presents another challenge. Often professional players have concerns about their relationships with family. It is difficult to receive the full understanding of parents in South Korean society, where school performance remains the top priority of every family. Coaches and team owners in South Korea are renowned for their training tactics in producing competitive professional esports players. While the coach holds responsibility for training a performing team, the team sponsor invests an average of US $15,000 to maintain a secondary league team in South Korea.

Mr D is an esports professor who heads one of the few esports departments in South Korea's higher education system. Professor D started his career in professional gaming and has experience with the first-person shooting genre and *Counter-Strike*. He later moved to lead esports teams as a coach. Because of his passion and expertise in esports, Mr D decided to stay in the esports profession. From there, he developed work opportunities in higher education. He has sometimes travelled between South Korea and China to manage new business in esports curricula. Mr D's expertise and history of personal participation in esports makes him popular among his students. Most of the students I interviewed during my fieldwork wanted to become professional esports players after graduation. In order to provide the required academic training, Mr D covers esports gaming skills, business management, and social media production in his university curriculum. Professor D's esports education has created a new aspect of higher education in South Korea, as his student teams can travel around the world to participate in tournaments and perform competitively to win awards in international tournaments.

Mr E is a passionate gamer who wanted to become a professional *Counter-Strike* player. However, he did not make it because of his age, having reached the normal retirement age of 25. Mr E then became an esports announcer for the TV channel OnGameNet. Because of his special broadcasting style, he started his own social media channels, which analyse gameplay and gather fan followers. Mr E is one of Korea's popular shoutcasters on platforms like YouTube and Twitch. The nature of the job depends on the esports seasons, where Mr E manages both his online streaming career and event commentary for e-stadium broadcasting as a freelancer. Mr E is a typical talent in the esports industry, someone who turns his gaming skills into emerging professions that are developing within the growing esports digital economy. There is an interesting self-growth process in Mr E's career development. His announcing style is loud, full of energy and somewhat dramatic and humorous. Mr E mentions that he has sometimes developed his shoutcasting style by referring to other successful Western shoutcasters on YouTube. Mr E sees his streaming career on Twitch as a kind of TV business. The career is challenging, as he has to stay on the top list of most-watched esports videos on websites such as AfreecaTV, where statistics on most-watched esports videos is public information. Mr E displays high enthusiasm for his career. However, the guarantee of earnings through his various social-media platforms still demands a great effort in order to produce, promote and monetise his videos. It has yet to become a fully rewarding job in Korea's expanding esports industry.

At the time when I was conducting fieldwork for this chapter in 2018, Ms F worked in the esports policy office of a provincial government near Seoul. Because the Ministry of Culture, Sport and Tourism supported provincial efforts to develop esports culture, Ms F had assumed charge of her province's esports budget. Her job included coordinating with local esports businesses to organise amateur esports tournaments. Ms F successfully branded the city by managing esports events and hosting an esports team. The daughter of a PC *bang* owner, Ms F grew up in the heart of South Korea's burgeoning esports culture. She was familiar with the details of the esports business as she had witnessed her mother's management of their family-owned PC *bangs* in her city. Since small-scale esports competitions often took place in PC *bangs*, Ms F had become an expert in maintaining the necessary networking to execute esports policy. The esports network in Ms F's city reflects a larger esports industry ecosystem that nurtured amateur esports in South Korea at that time. Many players who are interested in joining the esports industry can participate in jobs that relate to event planning and policy regulation. This approach toward monetisation, however, somehow relies on government budgets and can also reflect the unstable part of the industry's value-chain extension, as hosting esports tournaments and owning an esports team involve big budgets and high spending in order to see policy results.

Mr G and his peers are a group of young esports players I met during my fieldwork from 2017 to 2018. They are typical of many passionate esports gamers in Korea who are trying to enter the competitive esports industry. All the player interviewees were either in their late teens or early twenties. In South Korea, this age bracket is extremely important, as there is an emphasis on studying and preparing for college during this period. Unlike most college students in Korea, Mr G dropped out of school to become a full-time professional player. He played in the secondary esports league and lived with his teammates in an apartment near Seoul. Mr G enjoyed being a professional esports player. However, he admitted that the intensive daily training was sometimes stressful. Before he became a professional in esports, when he simply played for leisure in PC *bangs*, Mr G was consistently ranked among the top players in *League of Legends*. The professional lifestyle changed his performance at play. There were days that he was not satisfied with his own performance in competitions. The busy, tight schedule of training and competitions also forced Mr G to see his family less. Occasional small disagreements with teammates could also impact him psychologically when he competed. The most rewarding moments in Mr G's career has been the thrill of winning tournaments as well as enjoying interactions with his fans. The interview with Mr G offered a deeper understanding of the inner psychological state of young esports professional players, regarding their both personal and professional lives. Professional esports players are the drivers of the esports industry as they are responsible for the success of both the game corporations and the sponsors.

One of my other player interviewees, Mr H, was an emerging amateur player who shared Mr G's passion for esports. Mr H had just won a championship for the game *Clash Royale* in the high-school category when I met him at a tournament award ceremony in 2018. Despite his mastery in mobile gameplay, Mr H looked like a typical shy Korean high-school boy. The difference between Mr H and other normal high students was that most students his age were busy preparing for university. Mr H, however, had decided not to pursue a university education and hoped to enter professional esports after his graduation. The near future held a potential contract for him. At the age of 17, Mr H was one of the top players on the watch list for many local talent scouts and team coaches.

Esports in higher education is an emerging sector in the Korean esports industry. Many universities are in the process of setting up esports departments or offering esports-related courses. My conversations with a group of esports majors informed me of a variety of higher education esports training programmes that South Korea had pioneered. These students grew up as South Korea's gaming generation. They shared the memory of playing online games together with their childhood friends after school. The interviewed students agreed that esports education allowed them to improve skills to excel in competitive play. The curriculum covered teaching team values and strategies to support each other during matches. In addition, students also enjoyed broadcasting and public speaking courses, as they learnt how to stream effectively to online audiences over social media.

A 12-time *Tetris* champion in Korea, Mr I, told me that he did not regret exploring opportunities to start a new esports business. Mr I graduated with a degree in computer science from a prestigious university in Korea. He did not work in the IT industry. Instead, Mr I applied for a small government grant to start his own PR company. Mr I believed he was working on an innovative career path to promote esports culture. During his teenage years, he enjoyed playing competitively. Mr I's mother was initially worried about his 'passion' for gaming. But her concern was gradually relieved when she saw Mr I winning awards and prizes. At the time of the interview, Mr I's mother fully supported his plans to pursue a career in esports. Mr I shared that 'It is difficult to let go of something that you are very good at'. The words resonate with those of many other young players I met during my fieldwork in one of the most sophisticated gaming societies in the world.

Conclusion

Player-led innovation contributes to the development of the South Korean esports industry. There is still much effort needed to see the country's desired sustainability outcome, which will appear when the professionalisation process has fully transformed play into work. Since the esports infrastructure and its constituent hype began to take off in the early 2000s, the government has responded in a timely way to South Korea's growing esports fan culture. Considering the organisational setting, KESPA and IESF are two agencies that collaborate to promote the state's policy agenda to promote healthy esports in both national and international settings. KESPA is achieving its goal of standardising esports rules and monitoring the licensing system to manage professional players. Meanwhile, the IESF is achieving its goal of creating an agreed framework for maintaining player agreements that follow esports regulations and rules standardised by the organisation.

In subsequent follow-up fieldwork, I observed a value-chain phenomenon in the making of South Korea's esports ecosystem. When players move beyond their professional gaming careers, they utilise the resources they have accrued to create new growth in the already vibrant esports industry. This evidence of value chain creation in the making reveals a player-based, bottom-up approach to the search for paths of industrial development with the goal of monetisation.

The efforts of value chain creation can be seen in five industry areas affecting existing game businesses: publishing, sports, education, media and regulation. New job titles are created in five identified areas, and a comprehensive overview of all these activities confirms the leadership status of player-led innovation in the South Korean esports industry. These five areas all matter to the effort to build an ecosystem for healthy esports. Since professionalisation will inspire more audiences and fans to consume esports, an ecosystem to increase participation in the esports industry will then require careful planning to prioritise issues when discussing the agenda for making global esports policy.

The final statement of this chapter points to subsequent follow-up efforts to define the public sphere of the global esports ecosystem so that player welfare is placed closer to the centre of attention in future esports industry development. Most esports players start their careers very young and oftentimes have put off their studies in order to concentrate on developing into full-time esports players. A healthy esports ecosystem cannot avoid responsibility for the consequences of this. There should be an effort to build a healthy system that allows both easier entry into professional competitive gaming teams and more secure retirements for esports players. The attraction of esports should not be simply about player celebrity or winning games. Rather, it should be about playing esports as a basic cultural right for people who decide to participate and engage with other virtual players in esports culture. As the esports economy continues to organically grow, the public seeks to build a grassroots community that embraces the core values of esports while continuing to support players during the transformation from play to work.

Funding
This work was fully supported by a grant from Research Grant Council of Hong Kong Special Administrative Region (Project no. 14604215).

Notes
[1] The game is available to play at http://www.pyongyangracer.co/index.html.
[2] Names have been hidden to protect the privacy of the interviewees.

References

Ashcraft, B. (2018) 'The new definition of "pro gamer" in Japan', *Kotaku*, 2 February, Available from: https://www.kotaku.co.uk/2018/02/02/the-new-definition-of-pro-gamer-in-japan [Accessed 5 January 2021].

Bartle, R. (2004) *Designing Virtual Worlds*, Indianapolis, IN: New Riders.

Canning, S. and Betrus, A. (2017) 'The culture of deep learning in esports: An insider's perspective', *Education Technology*, 57(2): 65–9.

Casilli, A. (2017) 'Global digital culture/digital labour studies go global: Toward a digital decolonial turn', *International Journal of Communication*, 11: 3934–54.

CNBC (2015) 'North Korea's newest fad: The "Boy General" phone game', 9 December, Available from: https://www.cnbc.com/2015/12/09/north-koreas-newest-fad-boy-general-mobile-phone-game.html [Accessed 5 January 2021].

Eggert, C., Herrlich, M., Smeddinck, J., and Malaka, R. (2015) 'Classification of player roles in the team-based multi-player game *Dota 2*', 14th International Conference on Entertainment Computing (ICEC). Trondheim, Norway, pp 112–125, Available from: https://hal.inria.fr/hal-01758447/document [Accessed 5 January 2021].

eSportsflag (2020) 'Top esports teams from Korea', Available from: https://esportsflag.com/korea/teams [Accessed 5 January 2021].

Fuchs, C. (2013) 'Class and exploitation on the Internet', in T. Scholz (ed) *Digital Labour: The Internet as Playground and Factor*, New York: Routledge, pp 211–24.

Hoppe, D. (2019) 'The remarkable success of Japanese Esports Union (JeSU)', *Gamma Law*, Available from: https://gammalaw.com/the-remarkable-success-of-the-japanese-esports-union/ [Accessed 5 January 2021].

Huhh, J. (2009) 'The bang where Korean online gaming began: The culture and business of the PC bang in Korea', in L. Hjorth and D. Chan (eds) *Gaming Cultures and Place in Asia-Pacific*, London: Routledge, pp 102–16.

Huizinga, J. (1955) *Homo ludens: A Study of the Play Element in Culture*, Boston, MA: Beacon Press.

International Olympic Committee (2019) 'Declaration of the 8th Olympic summit', Available from: https://www.olympic.org/news/declaration-of-the-8th-olympic-summit [Accessed 5 January 2021].

Jin, D. (2010) *Korea's Online Gaming Empire*, Cambridge, MA: The MIT Press.

Kim, J. (2011) 'North Korean hackers hired to attack South Korea game network', *Reuters*, 4 August, Available from https://www.reuters.com/article/us-korea-north-hackers/north-korean-hackers-hired-to-attack-south-korea-game-network-idUSTRE77317720110804 [Accessed 5 January 2021].

Kucklich, J. (2005) 'Precarious labour: Modders and the digital games industry', *Fibreculture*, 5(1): 1–5.

Lund, A. (2014) 'Playing, gaming, working and labouring: Framing the concepts and relations', *TripleC*, 12: 735–801, Available from: https://www.triple-c.at/index.php/tripleC/article/view/536/625 [Accessed 5 January 2021].

Mozur, P. (2014) 'For South Korea, esports is national pastime', *New York Times*, 25 October, Available from: https://www.nytimes.com/2014/10/20/technology/league-of-legends-south-korea-epicenter-esports.html [Accessed 5 January 2021].

Nakamura, L. (2013) 'Don't hate the player, hate the game: The racialization of labour in the World of Warcraft', in T. Scholz (ed) *Digital Labour: The Internet as Playground and Factory*, New York: Routledge, pp 187–204.

Ratliff, J. (2015) *Integrating Video Game and Research and Practice in Library and Information Science*, Hershey, PA: IGI Global.

Rodriguez, H. (2006) 'The playful and the serious: An approximation to Huizinga's *Homo Ludens*', *Games Studies*, 6(1), Available from: http://gamestudies.org/0601/articles/rodriges [Accessed 5 January 2021].

Sholtz, T. (2013) 'Introduction: Why does digital labour now?', in T. Scholz (ed) *Digital Labour: The Internet as Playground and Factory*, New York: Routledge, pp 1–10.

Taylor, T.L. (2012) *Raising the Stakes: E-Sports and the Professionalization of Computer Gaming*, Cambridge, MA: The MIT Press.

Terranova, T. (2013) 'Free labour', in T. Scholz (ed) *Digital Labour: The Internet as Playground and Factory*, New York: Routledge, pp 33–57.

WHO (World Health Organization) (2018) 'Addictive behaviours: Gaming disorder', 14 September, Available from: https://www.who.int/features/qa/gaming-disorder/en/ [Accessed 5 January 2021].

Witkowski, E. and Manning , J. (2018) 'Player power: Networked career in esports and high performance game streaming practices', *Convergence*, 25(5–6): 963–9.

9

Representations of Play: Pachinko in Popular Media

Keiji Amano and Geoffrey Rockwell

These silver balls are you. They're your life itself.
(Ikiru, 1952)

How is play represented in Japan? For more than 60 years, the most popular leisure activity among the Japanese has been pachinko, so it is no surprise that pachinko has been represented in the arts from cinema to novels. Notable cinematic works include Ozu's *The Flavor of Green Tea Over Rice* (1952) and Kurosawa's *Ikiru* (1952). But what do these representations show us about the play in pachinko? Do the representations and culture of pachinko change over time? This chapter will look at a selection of Japanese representations of pachinko as a way to understand how pachinko becomes a metaphor for Japan.

Pachinko, as a game of chance, is one of the most popular games in Japan. The entire size of the leisure industry in Japan is about US$672 billion in 2019, and pachinko accounts for a third of that. However, it is rarely discussed in game studies in the West. For that matter, it is also ignored by the Japanese academic community, largely because it is an embarrassing form of gambling. Thus, the study of pachinko is not deemed to be serious. Media representations of the game then are one of the few ways to understand how the game is represented in Japanese culture.

In this chapter we will give an overview of how Japanese representations of pachinko changed and we will contextualise these changes in terms of the changing socio-economic environments. We will look at a selection of Japanese and Western representations of pachinko as a way to understand the history of the culture of this neighbourhood amusement. To do this we will begin with a short introduction to pachinko, and then look specifically at how pachinko was represented in three economic periods: (1) the 1950s and 1960s when the first parlour boom took place and the economy was taking off; (2) from the 1970s to the early 1990s, the period of the economic bubble; and (3) the latest pachinko related films after the economic bubble burst.

What is pachinko?

Any visitor to Japan will almost certainly notice the pachinko parlours. Pachinko is related to pinball in that the player fires steel balls that arch over a vertical playfield to then drop through pins and other obstacles. The goal of pachinko, however, is not to keep a ball in play like in pinball, but to get balls into special 'win pockets' or 'start pockets' that win the players more balls or start a video slot game. During the play, the players aim to fire balls so that they fall through the pattern of pins, gates, spinners, and chutes that bring the greatest likelihood of dropping the balls into the win pockets. They can do this by controlling the force with which the balls are launched. If the players are good then they will win more balls than the number they fire, leaving the pachinko parlour with prizes based on the number of balls they win.

To understand the cultural context of pachinko one needs to look at the chronotope (time and place) of play and how pachinko has been woven into the economy. In Japan it is illegal to gamble, but certain indirect forms of gambling like pachinko have evolved to be legal and visible. In fact, pachinko parlours are found at major city intersections occupying valuable real estate. Visitors who enter them are overwhelmed by the cascade of thousands of steel balls falling through pins. And yet, this game is ignored by scholars in game studies and Japanese studies, first because it is an almost exclusively Japanese phenomenon; and second because it is regarded as a disreputable form of gambling. At the same time, pachinko is so pervasive in Japan that the game no longer needs to be explained to the locals as it does in the West.

In the following section we examine media representations starting in the 1950s during the initial boom of parlours in which pachinko became widely popular. According to the industry association Zenkoku Yugigyo Kumiai Rengo Kai (1977), the number of pachinko parlours reached its peak in 1952 when there were 42,168 parlours. So, in the 1950s, pachinko as well as cinema became new everyday entertainment after World War II. Since then, pachinko has been a national pastime even though there has been government control to regulate 'gambling elements'.

Before turning to films and novels, a few words on the use of film and fiction as sources for this exploration of the history of playing pachinko. As Chapman (2011) and Deshpande (2004) point out, film can be used as historical or social documents, but like any source needs to be read critically. While there is a temptation to believe visual records like photographs or films directly represent reality or are even more 'real' than reality, one still needs to be aware of how films 'do more than simply reproduce the required object; they sharpen it, impose a style upon it, point out special features, make it vivid and decorative' (Chapman, 2011: 364). Films and novels do not just represent a phenomenon like pachinko, they also tell us about the attitudes towards the phenomenon of the creators in their time. In contrast to non-fictional works, films, novels and other fictional works have the additional advantage of showing us the game as played in a lived context, something that historical photographs and documentaries cannot show.

The Flavor of Green Tea over Rice (1952)

It is in this context that we look at how Yasujiro Ozu's *The Flavor of Green Tea Over Rice* depicted pachinko. Ozu, along with Kurosawa, is one of the greats of the classical Japanese studio (Russell, 2011). He made a number of films from silent ones in the 1920s to talkies up to the early 1960s, the most famous of which is probably *Tokyo Story* (1953). Many of his films do not have much of a plot but are 'home dramas' that portray the lives of ordinary people of those times. The title of *The Flavor of Green Tea Over Rice* already tells the audience that the film is about everyday life because green tea over rice (*ochazuke*) is a common dish, where hot tea is poured over cold rice as a way of using leftover rice for a quick bite. The movie shows an unhappy but well-off couple Taeko Satake (played by Michiyo Kogure) and Mokichi Satake (played by Shin Saburi) who are visited by a niece who is supposed to be meeting suitors for an arranged marriage. Leftover cold rice is a metaphor for the couple's marriage and the vivacious niece is the warm green tea. The niece's resistance to arranged marriage reanimates the couple's marriage just as warm green tea makes cold rice flavourful.

The movie repeatedly shows pachinko being discussed and played in a positive light over the course of the movie. Ozu was fond of everyday entertainments like pachinko and uses them as settings and metaphors. To begin with, pachinko frames the central issue of the movie, namely marriage of the main characters (Taeko and Mokichi Satake) and the marriage-to-be of their niece. In one of the earliest scenes, when Taeko and the niece visit a friend who owns a dress shop, they talk a little bit about pachinko and the friend says that being unmarried is the best time of life, the time when one can have fun. Taeko says, 'married life isn't easy [...] you will have no time to play'. Pachinko is represented as a novelty, a harmless pastime of youth, something fun that people have to set aside when they get married and become serious about life.

Later in the movie, Mokichi, the husband, who plays pachinko regularly, discovers that the owner of a pachinko parlour is someone he knew from the war and is invited over for a drink. Talking about pachinko as a new fad, Mokichi suggests that he sees no harm in the game, 'I know why it's the rage [...] It's fun'. The parlour owner replies, 'it shouldn't be regarded as fun [...] if so, the world won't improve [...]'. This dialogue shows how play and work are being negotiated in Ozu's work in post-war Japan.

The film also shows the business side of the game. At one point the parlour owner is shown leaning over the top of the cabinets when he is probably refilling balls and tending to the machines. He tells the players to try machine number four, likely the winner of that day.

The movie shows a form of play that is different from what is seen today. In the 1950s, players fed the balls one at a time into a hole in the machine, they did not turn a throttle to release tens of balls at once. Players would watch each ball fall through the pins, tracking their trajectories. They would hold the balls in the left hand and if they won, they would scoop them out of the tray with their hand and feed them back into the machine. Today, all the machines have electric ball launch systems so the technique of launching does not constitute an important part of the play. Players also manage a lot more balls per minute than in the old days, making pachinko more of a statistical game than one of skill.

Ozu shows a variety of experiences of playing pachinko as well. While the young rebellious niece is clearly having fun, Mokichi does not look like he is. He plays the game mechanically, killing time as he stays away from his wife, to whom he does not talk. His younger friend, on the other hand, seems to play seriously like a professional. Because that friend and the niece seem to both enjoy playing, Ozu hints that they both would enjoy being married, as they share the same pastime. In terms of screen time, they spend more time together playing pachinko than the niece spends on meeting potential suitors. This hints that she will choose the young man eventually.

Ikiru (1952)

Another cinematic example from the 1950s is Kurosawa's *Ikiru* (Life), in which pachinko is used as a metaphor for a frivolous approach to life. *Ikiru* follows an otherwise boring bureaucrat who looks for a more meaningful life after being diagnosed with cancer. At first, he tries drinking, then having sex and gambling at pachinko. The bureaucrat Watanabe is led astray by a flâneur, or as he calls himself, Watanabe's 'Mephistopheles'. This delightful devilish character compares pachinko to life. As we watch a ball dribbling down and bouncing off the nails, Mephistopheles says, 'Listen, these silver balls are you. They're your life itself. People strangle themselves in their daily lives. This machine sets them free. It's a vending machine for dreams and aspirations'.

Ironically, Watanabe later finds meaning in life by helping build a playground for children, another site of play. Where Ozu shows pachinko in everyday life, Kurosawa uses it as a metaphor for life, commenting on its playful randomness.

Foundry Town (1962)

Kirio Urayama's movie *Foundry Town* forms a nice contrast to *Green Tea Over Rice* and *Ikiru* as it shows a working-class view of pachinko. *Foundry Town* is about a depressed town outside Tokyo. The main character, Jun, is a young woman who struggles to pay for school. Her family is poor, and her father drinks away what salary he has until he loses his job. Jun ends up asking an ethnic Korean friend for an introduction to the pachinko parlour so she can work there part-time. At the time it would have been one of the few well-paid part-time jobs available to a young woman like Jun, but it would have also been considered a slightly disreputable job.

Part of what makes *Foundry Town* historically interesting is that there are scenes of Jun filling the trays of pachinko machines in the narrow corridors between machines, a process that has since been automated. Urayama showed how popular the game was in a working-class neighbourhood, even in the midst of a recession and labour strife.

Manual labour was no longer needed when the automatic ball supplier (invented in 1958) enabled operation without backroom workers. This changed the layouts of parlours by eliminating the corridors between machines that workers needed to access to the backs of the machines to add in ball supplies. This film also shows that until the 1990s, when video games threatened the parlour business, pachinko has always been considered a recession-proof business in Japan.

Another important commentary on society in *Foundry Town* is a subplot that shows the repatriation of ethnic Koreans in Japan to North Korea, which started in 1959 and peaked in 1962. The movie follows a young rascal who gets swept up in the repatriation, as his family decides to go back to North Korea. The pachinko business has been dominated by ethnic Koreans and there have been fears that profits made in pachinko parlours were being funnelled back to North Korea to fund North Korean nuclear ambitions that threatened Japan (see Chapter 6 about North Korean nuclear weapons). In short, pachinko is tied to the complexities of how Japan deals with a major ethnic minority. In the early 1960s, Koreans were encouraged to return to their homeland. But the emigration did not last once the horrendous living conditions in North Korea became known.

Films after Kurosawa and Ozu (1960s–1980s)

There were a number of pachinko-themed films after Kurosawa, Ozu, and Urayama. These included pachinko comedies in the 1960s and '70s during the boom of the pachinko and film industries. In the late 1980s pachinko became a gigantic, yet opaque industry. All the small parlours shown in the 1950s films closed down and gave way to large, expensive parlours. In films like *Black Rain* and *Thunderbolt* in the West, and the V-Cinema series in Japan, parlours were represented as establishments that have connections with gangs (*yakuza*). As Zahlten points out, pachinko-themed videos became a subgenre of V-Cinema in the early 1990s. They spoke to the anxieties of economic insecurity in the period after the economic bubble burst (2017: 185). These direct-to-video works often included extensive information about how to play specific pachinko machines woven into the drama.

Information about how to play machines was important because new technologies changed the stakes of pachinko games in the 1980s. Mechanical components such as the nails gave up space to screens where video slot reels would spin adding a statistical component to the game. When the stakes of getting balls into win pockets were higher, players became more serious. Playing pachinko was no longer an idle pastime, something that one might do standing around while waiting for the film to begin or avoiding an arranged date. Playing became a matter of gambling against the statistical odds, not playing against a mechanical machine. The pachinko machines became more complicated when electronic devices – 'black boxes' – became harder to conquer. This changed the way players play and think about the game. The lack of transparency led players to suspect that the management may have installed 'something unfair' in the machines or that profit went to supporting anti-social forces such as the *yakuza* or North Korea.

Western representations: *Lost in Translation* and *Black Rain*

Non-Japanese films and novels tend to show the pachinko parlour as a glittery setting that stands in, along with other visual tropes, for a neon and exotic Japan (see also Chapter 2 about the Western gaze of adult video games in Japan). Sophia Coppola's *Lost in Translation* (2003) treats Japan as the incomprehensible Other that can be safely mocked (King, 2005). Despite the film's title, she is not really interested in translating Japan, let alone understanding it. The pachinko parlour is just another flashing neon space in the night of Japan that serves as a backdrop to show alienation between human beings.

The story is about an ageing actor (played by Bill Murray) who arrives in Japan to shoot a commercial for a Japanese whiskey brand. He is lost at the end of his career. In Tokyo, he meets a young college graduate (played by Scarlett Johansson) who is also lost for not knowing what to do at the beginning of her career. Japan, and all the ways that it is deemed untranslatable, becomes almost like a third character standing in for the alienation of hypermodern life. Coppola expertly uses a variety of stereotypical caricatures of Japan for comic effect. She deploys various tropes in a discourse of 'technologies of recognition' (Shih, 2004). Viewers recognise the Japan-as-the-Orient in night scenes of Tokyo where the main characters run through a pachinko parlour. The recognition sets up the West as the recogniser and the Orient as the recognised, even though few Japanese would recognise themselves and their Japan in this movie. At no point was Coppola interested in Japan in and of itself as it recognises itself (see the Introduction about some scholars' objections to studying Asia as a region). As Day (2004) put it in a somewhat exaggerated critique of the movie in the *Guardian*, there is 'no scene where the Japanese are afforded a shred of dignity. The viewer is sledgehammered into laughing at these small, yellow people and their funny ways'.

In *Lost in Translation*, pachinko is used as a contrast with traditional Japan. Day (2004) pointed out that 'while shoe-horning every possible caricature of modern Japan into her movie, Coppola is respectful of ancient Japan. It is depicted approvingly, though ancient traditions have very little to do with the contemporary Japanese'. In this movie, Coppola showed that it is fine for a Westerner like her to like flower arranging and Buddhist monks chanting, but she mocked contemporary Japanese culture perhaps because it is a threat to the American ownership of modernism. There is a fear that Japanese modernity will replace the American one. To suppress this fear, Japanese modernism therefore has to be recognised as ridiculous, without any real progressive future. In other words, Japanese modernity is shown to be infertile, childless, and productless in *Lost in Translation*. Pachinko is the epitome of this alien productless-ness, and is therefore easily dismissed as a technological dead end. Even the Japanese recognise that it is a 'Galápagos industry' that cannot be exported outside Japan (In Japanese, a Galápagos industry is one that has evolved alone on the Galápagos Islands, producing goods that do not have a market overseas). As Day (2004) puts it,

> The US is an empire, and from history we know that empires need to demonise others to perpetuate their own sense of superiority. Hollywood, so American mythology has it, is the factory of dreams. It is also the handmaiden to perpetuating the belief of the superiority of US cultural values over all others and, at times, to whitewashing history. (para. 8)

How might Japanese forms of entertainment be taken seriously if they are indeed the future of gaming? How will these forms, including the particular one of pachinko, shift the attention away from Western trends? These questions raise the issue of techno-orientalism reflected in Ridley Scott's *Black Rain* (1989). We will explore how Scott used a more technologically oriented representation of modern Japan and pachinko. The representation of Japan in this movie looks a lot like the dystopic future shown in *Blade Runner* (1982). Japan is a country at night, dark with lots of neon and steam venting dramatically. In *Black Rain*, the enigmatic tall woman looks like Rachael from *Blade Runner*, the experimental replicant with memories. However, *Black Rain* takes place during the height of Japanese economic power, unlike the messy and corrupt Los Angeles in *Blade Runner*. The Japan in *Black Rain* is vibrant with large-scale manufacturing plants. One key scene takes place in a foundry that visually recalls one of the scenes of the Nostromo in *Alien*. Ironically, 1989 was the year that marked the end of the economic boom. A draft script of *Black Rain* describes the first pachinko scene:

> INT. PACHINKO PARLOR – DAY
> The largest in Tokyo. Endless rows of men sitting in front of the machines (horizontal pinball games) furiously punching the flippers. The NOISE from the metal balls is deafening. (Bolotin and Lewis, 1987)

It is unsurprising that pachinko parlours would serve as a setting for *Black Rain* because by the late 1980s pachinko parlours served repeatedly as a visual trope for Japan as the cyberpunk Other. To Japanese designers, pachinko parlours are designed to stand out because they need to be visible to customers. They are designed to showcase 'the building itself as a sign for the parlour' (Verena, 2008). These spaces become signs of potential play to the Japanese in Japan; as well as signs of techno-Japan, both attractive and alien, whether in films or novels. In *Black Rain*, the representation of the pachinko parlour is, however, different. What is shown is the dingy back offices during the daytime rather than the usual shiny play spaces at night. The dingy offices become the main location for action. In this way, Scott is closer to how pachinko might be represented by a Japanese director, not an entertainment venue parlour with machines, but as a business with potential ties to crime.

Kugi-shi (1978) and *Tokyo-ga* (1985)

The 1980s saw a shift from mechanical to electrical pachinko. This transition was documented in the 1978 Japanese short novel *Kugi-shi* (Nail doctor) by Giichi Fujimoto and a documentary *Tokyo-ga* by Wim Wenders. *In Kugi-shi* and *Tokyo-ga*, the main character is the *kugi-shi* (the nail doctor).[1] Fujimoto's story *Kugi-shi* is a duel between skills, setting the pachi-pro against the *kugi-shi* who is the professional adjuster of nails on the pachinko machine. They arrange and fix the angle of the nails on mechanical pachinko machines to control the possible flow of balls and hence the payout. Pachi-pros who have different techniques try to beat Kugi-shi's nail setting so that they can win the game. In this story, the nail doctors not only played an important role in the parlours, but were regarded as artisans by parlour staff, customers, and even pachi-pros.

In the era of mechanical pachinko, pachinko was both a game of luck and one of skills/technique, not a statistical game like today. Some played for fun, some killed time, but the pachi-pros aimed to make money against the parlour. To do so they had to find machines with nails arranged in ways that maximised their chances of bouncing the balls into the win pockets. This meant playing against the nail doctors, whose job was to straighten out the nails. Skilled pros knew what to check: nail layout, condition of the launch handle, the personality of the parlour manager and nail doctors, discharge history as recent as that of the previous day, and even the weather because humidity affected the shape and angle of the wooden backboard.

Parlours defended themselves against the roving pros by sending in the nail doctors to straighten pins after hours. A *kugi-shi* in Fujimoto's short novel says:

> because it is launched from the flipper and fingertips, it is natural that it goes in the wrong direction. It's much like life. You sometimes face a little nail and it will bring you to wander somewhere you don't want to go. But he leads the balls to the same course in 35% probability. (Fujimoto, 1978: 14)

Pachinko captures the randomness of life; pachinko is used again, as in *Ikiru,* as a metaphor for life.

The *kugi-shi* is also depicted in Wim Wenders' documentary *Tokyo-Ga,* a film about the director Yasujiro Ozu. The documentary shows a moment of transition from an era of mechanical pachinko and the post-war cinema of Ozu, Urayama, and Kurosawa to that of increasingly electronic machines where pachinko is an idiosyncratic version of a slot machine where players' skill is irrelevant to the game.

Wenders' documentary, like Barthes' *Empire of Signs* (1982), is a sympathetic exploration of Tokyo and modern Japanese culture that does not get lost the way that Coppola does. Like Barthes' work, Wenders' documentary includes a meditation on pachinko. He tries to understand Ozu and Japan through pachinko and settles on the game as an understandable way of forgetting in the face of both personal and national traumas. He scales up the way people might play to kill time, suggesting the game is a nation playing to forget about the traumas brought by the war. Even though this is not an original point about pachinko, it still brings the viewers back to Ozu in which play is situated in the face of war and work.

As Wenders' narrator puts it:

Late into that night,
And then late into the following nights,
I lost myself in one of the many pachinko parlours.,
And the deafening noise,
Where you sit in front of your machine,
One player among many,
Yet for that reason all the more alone,
And watch the countless metal balls dance between the nails
On their way out, once in a while, to a winning game.

This game induces a kind of hypnosis,
A strange feeling of happiness.
Winning is hardly important,
But time passes,
You lose touch with yourself for a while,
And merge with the machine,
And perhaps you forget
What you always wanted to forget.

This game first appeared after the last war
When the Japanese people had a national trauma to forget.

Wenders' film also shows the *kugi-shi* (nail doctor) at a time when the profession was becoming extinct even though the character did not realise it at that time. In the film, the director narrated: 'only the *kugi-shi*, the nailman, is at work. Tomorrow, all the balls run different courses, and the machine that made you a winner today, will only make you a desperate loser tomorrow.'

Gold Rush (1998)

Miri Yu's novel *Gold Rush* (1998) was published after the bursting of the economic bubble in Japan. It also presents pachinko as a metaphor for Japan, but a very different Japan from that represented in the works above. *Gold Rush* is about the 14-year-old spoiled son of a wealthy pachinko business owner in a seedy part of Yokohama. The son ends up killing his father and trying to take over the business. A lot could be said about this novel which, like any good work, defies easy interpretation, but in the context of this chapter, we will focus on a reading of the place of pachinko.

To begin, the author Miri Yu is a *Zainichi* (Japanese-Korean). Her father fixed pachinko machines so she was familiar with the business of pachinko. Japanese readers would know that she was Korean just from her name. They would also associate pachinko with the *Zainichi*, justifiably since the majority of pachinko parlours are owned by ethnic Koreans. But Yu surprises her Japanese readers by using pachinko not as a metaphor for the *Zainichi*, but as a metaphor for a valueless Japan at the end of the economic boom. Yu turns a game that is considered the business of the *Zainichi* back onto Japan as a dark reflection of its emptiness after the bust. Where a reader might have expected pachinko to be used to stand in for the ethnic Korean community, as it is in the novel *Pachinko* by Korean-American author Min Jin Lee (2017), it is used by Yu to reflect on what Japan had become once it was no longer as economically powerful after the real estate bubble burst in 1991.

Second, Miri Yu's *Gold Rush* references *The Temple of the Golden Pavilion* by Yukio Mishima (1956). *The Temple of the Golden Pavilion* is an earlier and similarly dark depiction of a Japan without any values, though Mishima is depicting a Japan during the American Occupation after World War II. The title *Gold Rush* references that of Mishima's novel, suggesting Yu is responding to Mishima's literary novel on the aimlessness and lack of values of Japan with a novel about a parallel lack of values after the economic bust. In this interpretation, the readers are confronted with the paradox of Yu turning Mishima's traditionalist critique back onto itself. *Gold Rush* was written for the Japanese and she represents Japan back as something they thought was not Japanese, namely, Korean pachinko culture. Where Mishima bemoans the loss of traditional Japan after Americans brought in Western capitalism during the occupation, Yu shows a Japan born of both empty traditions and modern capitalism. She shows the traditional culture to be just as empty as the productless business of pachinko. The violent son is abandoned by both his wealthy businessman father who has no time to raise him and by his mother who abandons her children to pursue a pure life, free of consumerism. Even the weapons of choice symbolise the different values of Japan: The son kills the dog with a gold club that symbolises Japan's business culture; he then kills his father with a samurai sword that symbolises the traditional Japanese culture. Perhaps not coincidently, the sword also reminds us how Mishima ended his life committing ritual suicide.

Other notable works in the post-economic bubble era

Other post-economic bubble works are not as dramatic as *Gold Rush*. As mentioned earlier in this chapter, in the early 1990s many low-budget pachinko direct-to-video films (V-Cinema) were produced and many of them pictured pachinko in connection with anti-social forces. During this period, the average price of the stock market index for the Tokyo Stock Exchange (N225) fell dramatically, while the pachinko-playing population remained stable and the market size of the pachinko industry actually increased. Pachinko appeared to be a recession-proof industry.

However, the pachinko business changed in the late 1990s: Although the market size was still increasing, the number of players began to drop, as the unemployment rate for young people rose. Similarly, the number of films and novels that have pachinko as subjects decreased. Works produced since the late 1990s use pachinko as the setting for those left behind by the economy, especially unemployed young people.

In contemporary Japan, there is a talk of an 'invisible caste' left behind by an excessively competitive society. Nobuyuki Fukumoto's popular manga *Kaiji* (1996) is a good example. Kaiji, the eponymous hero, is at the bottom of this invisible caste system – a loser who is forced to gamble to survive. In one episode, he has to fight against a monster pachinko machine set in an underground casino. In the movie version of the manga, this monster machine, like the post-bubble economy, is hard to beat as the odds are terrible and the stakes are high. Pachinko becomes the site of a cruel duel between the left behind and the machine, where the left-behind strive for economic survival.

Conclusion

As Marshall McLuhan wrote, 'the games of a people reveal a great deal about them. Games are a sort of artificial paradise like Disneyland, or some Utopian vision by which we interpret and complete the meaning of our daily lives' (2003: 319). The tactics, experience and culture of playing pachinko has changed over time and these changes in media representations reveal how Japan is seen at play.

In the 1950s, when pachinko appeared as a new national pastime, film directors like Ozu and Kurosawa presented pachinko as a novel entertainment and a metaphor for life, especially in its randomness. From the 1960s to the 1980s the representations diversified. Sometimes, pachinko was used as a setting to show economic conditions, as well as to depict Japan's significant Korean ethnic minority. As the machines became electronic, the stories changed. Players were shown outsmarting the nail doctor, the local parlour, and/or *yakuza*. For the casual players of the mechanical age, pachinko was a game with some possibility of winning if you could read the nails and control the launch of balls. The stakes were low, so players would not lose too much money. However, they would not win that much either. Casual players could develop their own intuitive mechanical tactics; even if they did not work, and in that find repetitive play to pass the time.

By the 1990s, players no longer competed against the nail doctor or the local parlour. With digital machines, players now gambled with complicated odds and the stakes were higher. Casual play became less common, as players could lose or win more than they used to. The frivolous fun in Ozu and Kurosawa is forever gone; players now spend long hours seated in a parlour as if they were at work. Pachinko has become a symbol for and the victim of the uncertainties of post-bubble Japan. Like the ageing players, pachinko is a game left behind, as youths now play on smartphones or video game consoles rather than in parlours.

Note

[1] Another work that is not analysed here but also tells the story about the fights between pachi-pros (pachinko professionals) and a nail doctor in the 1960s and '70s is Jiro Gyu's manga *Kughi-shi Sabuyan* (1971).

Filmography

Alien (1979) Scott, R. (director), 20th Century Fox.
Black Rain (1989) Scott, R. (director), Paramount Pictures.
Blade Runner (1982) Scott, R. (director), Warner Bros.
The Flavor of Green Tea over Rice (1952) Ozu, Y. (director), Tokyo: Shochiku.
Foundry Town (1962) Urayama, K. (director), Tokyo: Nikkatsu.
Ikiru (1952) Kurosawa, A. (director), Tokyo: Toho.
Lost in Translation (2003) Coppola, S. (director), American Zoetrope.
Thunderbolt (1995) Chan, G. (director), Golden Harvest.
Tokyo-ga (1985) Wenders, W. (director), Chris Sievernich Filmproduktion.
Tokyo Story (1953) Ozu, Y. (director), Tokyo: Shochiku.

References

Barthes, R. (1982) *The Empire of Signs* (R. Howard, trans), New York: Hill and Wang.
Bolotin, C. and Lewis, W. (1987) 'Black Rain draft script', *Screen Plays for You*, Available from: https://sfy.ru/?script=black_rain_ds [Accessed 5 January 2021].
Chapman, J. (2011) 'Researching film and history: Sources, methods and approaches', in E. Margolis and L. Pauwels (eds) *The SAGE Handbook of Visual Research Methods*, London: Sage, pp 359–71.

Day, K. (2004) 'Totally lost in translation', *The Guardian*, 24 January, Available from: https://www.theguardian.com/world/2004/jan/24/japan.film [Accessed 5 January 2021].

Deshpande, A. (2004) 'Films as historical sources or alternative history', *Economic and Political Weekly*, 39(40): 4455–9.

Fujimoto, G. (1978) *Kugi-Shi* [Nail doctor], Tokyo: Kadokawa.

Fukumoto, N. (1996) *Kaiji*, Tokyo: Kodansha.

Gyu, J. (1971) *Kugi-shi Sabuyan*, Tokyo: Kodansha.

King, H. (2005) 'Lost in translation', *Film Quarterly*, 59(1): 45–8.

Lee, M.J. (2017) *Pachinko*, London: Apollo.

McLuhan, M. (2003) *Understanding Media: The Extensions of Man*, Berkeley, CA: Gingko Press.

Mishima, Y. (1956) *Kinkakuji* [The Temple of Golden Pavilion], Tokyo: Shincho-Sha.

Russell, C. (2011) *Classic Japanese Cinema Revisited*, New York: Continuum.

Shih, S. (2004) 'Global literature and the technologies of recognition', *PMLA*, 119(1): 16–30.

Verena (2008) 'Tokyo odyssey: Pachinko parlour glitz', (N. Yamane, trans), *PingMag*, 12 September, Available from: http://web.archive.org/web/20080915000538/http:/pingmag.jp/2008/09/12/tokyo-odyssey-pachinko-parlour-glitz/ [Accessed 6 January 2021].

Yu, M. (2002) *Gold Rush* (S. Snyder, trans), New York: Welcome Rain.

Zahlten, A. (2017) *The End of Japanese Cinema: Industrial Genres, National Times and Media Ecologies*, Durham, NC: Duke University Press.

Zenkoku Yugigyo Kumiai Rengo Kai (1977) *Zenyuren (Kyo) 25 nen shi* [The 25-Year History of the All Japan Game Business Cooperative Federation], Tokyo: Zenkoku Yugigyo Kumiai Rengo Kai.

10

The Work of Care in the Age of Feeling Machines

Shawn Bender

By the turn of the millennium, it was clear that Japan had a problem. Numbers that had risen steadily in the past now seemed likely to plateau or, worse, to decline. The situation was without precedent, at least since the good days of the 1960s. Its cause seemed obvious enough: The country faced a crisis of reproduction. Conditions that had held for much of the last half-century no longer did. Ten years of middling economic growth – the so-called 'lost decade' – had taken a toll. Vitality had slackened. Industrial output had slowed. A future of predictable growth had given way to a haze of doubt, uncertainty, and risk. Immigrant labour would only exacerbate the problem. Experts contrived a solution as unprecedented as the problem. It targeted the very heart of society itself: home, family, and community. Without attending to the more intimate spaces of social life and grappling with the inherent unpredictability, few believed there would be much chance of success.

★ ★ ★

Scholars of Japan have become familiar with such dire warnings about the country's population trends. It's not surprising why. The nation's birth rates are among the lowest in the world and have been for years. The proportion of seniors in Japan remains the world's largest and is projected to increase even more in the future. Consciousness of low birth rates and high longevity is so prevalent in Japan that demographic change has become a lens through which many see the social body of the nation itself, as a '*shōshi kōreika shakai*' (low fertility, ageing society). In 2014 a Pew Research Center survey found that a whopping 87 per cent of those surveyed in Japan perceived the country's growing number of seniors to be a problem (Pew Research Center, 2014).

But the discourse of declinism in Japan is not limited to demographic concerns alone. The gloomy scenario with which this chapter begins does not express worries about human reproduction. It reflects concerns about flagging demand for Japan's population of non-human industrial robots.

For much of the twentieth century, Japan led the world in the application of robots in automobile manufacturing and other industries. At the turn of the twenty-first century, however, experts in the robotics industry feared that demand for their machines, both domestically and internationally, had either already peaked or would do so soon (JARA and JMF, 2001; Yokoyama, 2004).[1] Robot makers and robot users had built up symbiotic relationships over years of steady demand, whereby multipurpose machines were tailored to the specific needs of factory owners. These relationships helped ensure the efficiency of assembly lines and the safety of humans working in them. Lower demand for industrial robots meant that established relationships between the corporate makers of industrial robots and their corporate users would not be (re)produced as reliably as before. Makers would need to find markets outside industry. Robots, they came to believe, would have to roam free of factories and begin '*kyōson*' (co-existing) with humans – in communities, in businesses, and in homes. In effect, the needs of people in these places would become the 'problems' that a new generation of robotic devices would help solve, much as they had once done for industry. The era of robot–human co-existence would render everyday environments anew as fields for technological experimentation – everyday life would become a platform for innovation.

The post-industrial turn in the robotics industry reflects a broader shift in the Japanese economy that has intensified since the start of the twenty-first century. Observers have noted how the relatively stable socio-economic institutions of late-twentieth century Japan began to break down beginning with the collapse of the 'bubble' economy in the early 1990s. This continued into the twenty-first century, as Japan faced the dual challenges of globalisation and economic malaise (Yamada, 2004; Genda, 2005; Kosugi, 2008; Brinton, 2011; Allison, 2013; Koch, 2016). The stability of school-to-work pathways, lifetime employment (for male employees of large companies), and seniority-based promotions gradually gave way to demands for a more flexible supply of labour, merit-based promotion, a diversification of occupational preferences among youths, higher rates of unemployment, and growing levels of economic inequality (see also Chapter 1 about flexible employment among South Korean females). While the changing economy and accompanying neoliberal economic reforms created opportunities for ambitious men and women who could rely on newly fashioned 'commodified selves' (Takeyama, 2016; see also Lukács, 2015; 2020), concern arose about higher levels of social isolation and disconnection. In 2009, news agencies reported that the number of ' *kodokushi*' (lonely deaths) – a term that refers to elderly individuals who die alone – had reached 32,000 (Allison, 2015: 131). The following year a documentary televised on NHK, Japan's public broadcaster, introduced the term '*muen shakai*' (the society of no-relation) to characterise the profound dissolution of everyday social ties that had produced such a grim statistic (NHK Special, 2010). Along with contemporaneous news that over 200,000 centenarians who were thought to be alive were actually dead or missing, the 'society of no-relation' referenced a rupture in intergenerational communication, 'the general indifference that people feel toward elders and, perhaps, the general indifference that elders feel toward society' (Nozawa, 2015: 376). Japan was not just 'ageing' and deindustrialising; citizens seemed to be withdrawing from each other and receding from the social institutions that had formerly rooted identity and community.

It was in this fraught social context in late 2010 that I began ethnographic fieldwork in Japan with a group engaged in a new form of care for populations of older adults. In sessions of what they call 'robot therapy', the group uses robots in activities meant to ease the psychological and behavioural symptoms of dementia among older adults in nursing homes. The robots they use are not the intimidating behemoths of industry one might at first imagine. Instead, they employ consumer-oriented devices that are smaller, friendlier-looking, and designed for entertainment and interaction. These robots are engineered to elicit an affective connection with human users and to sense shifts in their emotions and intentions. They are what I call 'feeling machines': digital devices made to be felt and programmed to feel. These range from toys like Tamagotchi to more sophisticated animaloid robots such as Sony's AIBO robot dog, Omron's NeCoRo robot cat, and Intelligent Systems' Paro robot seal.

In 2010 it seemed a foregone conclusion to many whom I met that Japan would have to rely on the physical and affective labour (Hardt,

1999) of machines to make up for a projected shortage of professional care workers in the coming aged society. Indeed, using these robots with seniors placed the group on the vanguard of applying to elder care what would come to be known as '*kaigo robotto*' (care robots).[2] The first local government project supporting the adoption of care robots in elder care, the *kaigo-iryō bunya robotto fukyū suishin jigyō* (Nursing Care Robot Promotion Project) in Kanagawa Prefecture, concluded a few months before I began work with the group. In 2012 the Ministry of Health, Labor, and Welfare launched a national effort to promote care robotics, the *robotto kaigo-kiki kaihatsu dōnyū sokushin jigyō* (Robotic Devices for Nursing Care Project), which remains active as of the time of writing this chapter (Robotic Care Devices Portal, 2019).

These public initiatives tend to present the emergence of care robotics for elder care as a natural, if inevitable, response to the labour needs of an ageing society. This assumption is shared by other published studies of care robots in Japanese nursing homes (Tamura et al, 2004; Wada et al, 2004; Ishiguro, 2017; Brucksch and Schultz, 2018; Wright, 2018). When I first encountered the group, I, too, assumed that they were interested in exploring how autonomous robots might perform the labour of care without human intervention.

I learnt quickly that the idea of robots as substitute labourers held little purchase among members of the robot therapy group. As engineers-turned-educators, few of them believed the hype disseminated by government officials and media organisations. They were keenly aware of the limitations of robot technology, and they did not see robots as viable replacements for human care work now or in the future. Instead, they considered robots to be tools to care with, supplementing rather than substituting human affective labour. In sessions, they worked with a multitude of robot 'colleagues' to create periods of activity that not only provided a break from the monotony of nursing home life, but also created opportunities for social interaction and collective engagement. Sessions were designed to reduce feelings of alienation and isolation among residents by restoring, albeit briefly, a sense of connection and community. Moments of connection and community like these were believed to have a therapeutic effect.

Yet, as much as the group was committed to facilitating dementia care, I found that they were driven even more powerfully by a passion for exploring how robotics technology might be applied in everyday life. For them, elder care offered just the kind of everyday site of investigation that leaders of the robot industry had proposed years earlier. They saw their investigations as ongoing and open-ended – the kind of improvisational approach to using technologies in care that scholars of science and technological studies have affectionately called 'tinkering with care' (Mol et al, 2010). Importantly, their tinkering did not end at the conclusion of robot therapy sessions. It resonated after in conference presentations, scholarly publications, student projects, and in exchanges with the makers of robots themselves. The group hardly saw its activities as confined to dementia care. They viewed themselves as active participants in a larger conversation among roboticists about the relevance of robotics technology in Japan, a country buffeted by the forces of de-industrialisation, demographic change, and social dislocation.

In this chapter, I describe how the group came to use robots with older adults and provide an ethnographic account of how they care using robots in sessions of robot therapy. In these sessions, the group uses robots to create a stage of engagement that promotes sociality as a form of care. Sessions effectively transform robots into technological platforms for social interaction just as they transform nursing care into a real-world platform for robot experimentation. I suggest further that the group's interest in effecting sociality addresses more than the perceived needs of patients with dementia; it also signals the reparative potential of feeling machines in Japan's 'society of no-relation'. The group's manipulation of robotics technology helps to create the very conditions for sociality through communication, what the linguist Roman Jakobson calls language's 'phatic function' (Jakobson, 1990). This discussion of human–robot engagement begins, oddly enough, with the substitution of feeling machines for cuddly pets.

From animals to animaloids

In 1999, one of the founders of the robot therapy group, a man I call Prof Matsuda,[3] retired from a career as an engineer at a large Japanese electronics firm to assume a teaching position at a university in Tochigi Prefecture. His assumption of a new career after retirement is not unusual. Given that most large firms in Japan mandate that employees retire in their 50s, many former salaried workers feel that they still have productive years ahead of them. For Prof Matsuda, this meant a new life in the academy. He taught in Tochigi Prefecture for five years and then transferred to another university in Tsukuba City, where he worked until 2019, when he retired again, this time for good.

Soon after assuming his position in Tochigi, he happened to see an NHK report about a psychiatrist, Dr Yamaoka, who had started using robots for paediatric care in hospitals. Before starting to use robots, Dr Yamaoka had been an enthusiastic advocate of Animal

Assisted Activity (AAA) and Animal Assisted Therapy (AAT). The two are closely related but AAA involves interaction with animals and is loosely organised, while AAT uses animals as part of a goal-directed treatment plan. Promoting active, tactile interaction with animals is fundamental to the success of AAA and AAT alike. The goal is to achieve a therapeutic effect by generating affects of happiness, calm, and comfort.

Dr Yamaoka told me in an interview that in the course of working with AAA and AAT in hospital care, he began to encounter multiple challenges. First, animals – in this case, usually cats or dogs – need to be fed regularly and cared for continuously, which can prove challenging for hospitals that choose to offer animal therapy. Even if hospitals outsource the service, therapists need to secure the understanding of the participating hospital and of the families of all participating patients. Patients might be scared of animals or have allergies, which limits the population that can engage in such therapy. Additionally, animals can get stressed or tired through repeated 'work' with patients who might also unintentionally treat them in a rough or injurious manner. Perhaps most surprising, Yamaoka encountered what he called 'the discriminatory attitude toward animals in Japan'. Japanese people, he told me, tend to think of animals as 'dirty' or 'dangerous'. This leads to a somewhat irrational resistance to animals on the part of some health professionals, but it also expresses a perhaps more understandable concern that animals might introduce pathogens into environments where hygiene is of paramount importance.

Frustrated by these limitations, Dr Yamaoka tried instead using the then recently released Sony AIBO robot dog with hospitalised paediatric patients in precisely the same manner that he had previously used animals.[4] He encountered much less resistance from hospital staff. He discovered further that interactions with robot pets – AIBO at first and later the robot cat NeCoRo – provided ' *shigeki*' (stimulation) for patients and helped '*antei saseru*' (relax) them much as interactions with animals had done. Using a buzzword of the early 2000s in Japan, members of the Robot Therapy Group (RAT) described the salutary effect of both animals and robots with the term '*iyashi*' – a word used popularly to refer to stress relief, relaxation, and restorative feelings of calm. Often this term is used synonymously with the transliterated English word 'healing' (*hiiringu*).[5] This is the kind of everyday 'healing' provided by services like spa visits, saunas, massages, and so on that one might visit to relieve the stress of everyday life. Both robots and animals, as it were, helped relieve the anxiety of hospital stays and promote healing-as-*iyashi*.

The NHK news report that Prof Matsuda saw described Dr Yamoka's efforts. It also noted that Dr Yamaoka was scheduled to present at a conference on human–animal relations. Prof Matsuda quickly registered for the conference and attended Dr Yamaoka's presentation. The two men struck up a working relationship and started a collaboration that continues as of this writing. In 2002, together with interested professors of engineering in the Tokyo area, the two founded an Animal Therapy/Robot Therapy Research Group under the auspices of the Society of Instrument and Control Engineers (SICE). The group became a formal chapter of SICE the following year.

The group's founding documents reflect the early twenty-first century attitude to robotics outlined earlier. Its mission statement recognises a 'new wave' of robotics technology, one characterised by the arrival of post-industrial '*pāsonaru robotto*' (personal robots) that will co-exist with humans outside of factories. This optimistic depiction of technological progress is followed quickly (but without an obvious logical connection) by the recognition that Japan will soon encounter a '*chō-kōreika shakai wo chokumen shite-iru*' (super-aged society). The statement emphasises how personal robots could provide value for older adults in such a context, particularly by enhancing the 'quality' of their everyday life by facilitating 'communication' with those around them (RAT/AAT Research Group, 2002). In explaining why, the group contrasts the purpose of newer post-industrial robots with their antecedent industrial robots. Whereas industrial robots sought to increase the speed and efficiency of production, personal robots could instead enhance feelings of comfort and satisfaction. The change involved a shift of emphasis from the quantities of robotic production to the immaterial qualities of everyday life – that is, from the creation of objects to the elicitation of affects, particularly among older adults. Given the social problems of the 'ageing society', the group aimed to 'pursue the application of robots to nursing care and the welfare' of older adults (RAT/AAT Research Group, 2002).

Crucially, the effect of their activities would not be limited to care alone. Ultimately, they sought to turn these affects back into objects. They aimed to explore the possibilities of robots in care, to learn about what kinds of robots were most effective, to see what kinds of functionality were most useful – in short, to discover what value personal robots could bring to society. In so doing, in their language and in their aims, they reflect contemporary thinking among elites in the robotics industry who saw everyday life as a new, untapped market opportunity. Furthermore, without using the language of 'platform' specifically, they perceived the benefits that robots might bring to care and that sites of care might offer to technological innovation.

Robot Therapy for dementia care

The robot therapy group visits nursing homes periodically to offer sessions of robot therapy such as the sessions of animal therapy that Dr Yoneoka conducted. This makes them not altogether different from other groups that visit facilities to help residents get exercise, to cut hair, to perform comedy, or to do music or art therapy. Most elder-care facilities in Japan – indeed, worldwide – take advantage of such services when they are available. They are as much a part of 'good care' as the daily provision of meals and lodging. What distinguishes the robot therapy group from the typical itinerant care provider is that they are both part-time caregivers and full-time university educators who conduct sessions of robot therapy together with their undergraduate students. Working with these students, they track the results of therapeutic sessions, converting acts of care into quantitative data for analysis, for student projects, and for scholarly publication.

Yet, of course, the primary aim of their sessions is to help treat the symptoms of dementia among older adults in institutional care. The Alzheimer's Association defines dementia as a 'general term for a decline in mental ability severe enough to interfere with daily life' (Alzheimer's Association, 2018). Dementia is not synonymous with Alzheimer's disease (AD) but approximately 60 to 80 per cent of all cases of dementia stem from Alzheimer's. As the medical anthropologist Margaret Lock writes in her book *The Alzheimer Conundrum*, the cause of dementia has never been definitely proved (Lock, 2013: 3). A cure has similarly eluded the grasp of clinical medicine. The symptoms of people with dementia progress gradually and sometimes dramatically, and, tragically, they never improve. People with dementia become increasingly dependent to a level few families can manage. Many live out their days in institutional care.

Any romance that one might harbour about the miraculous capabilities of cutting-edge robotics dissipates quickly when one is confronted by the realities of caring for people with dementia. Even after repeated observation, I had difficulty reconciling popular discourse about futuristic technologies of care and the everyday challenges that care workers and recipients alike face in managing the symptoms of dementia. Responses to everyday rituals of waking, bathing, getting dressed, eating, and socialising can vary wildly, sometimes violently so. Such variation is undoubtedly due largely to the impact of neurophysiological pathology. But research from the 1990s has emphasised the role that social interaction can play in accelerating or arresting the progression of disease and in modulating the expression of symptoms (see Kitwood, 1990; 1997). It was in the facilitation of sociality where the robot therapy believed its technologies could make a valuable contribution to institutional care.

Robot Therapy as performance

I met Prof Matsuda in late 2010 when he was in his fifth year of doing robot therapy at a nursing home in Saitama Prefecture that I call Leisure Time. The nursing home is managed by his friend and former Waseda University classmate Okada-san. After Okada-san retired from a career as an engineer at another large Japanese company, he began helping his relatives manage the front office at Leisure Time. The two men stayed in touch over their careers, regularly exchanging customary New Year's Cards. When Okada-san learnt that Prof Matsuda had taken a teaching position in Tsukuba city, he wrote that he heard that Paro, the baby seal robot, had just been made available for sale there. This led to a conversation about Prof Matsuda's engagement in robot therapy and subsequently to regular visits of the group to Leisure Time (now numbering over 100 and counting).

Prof Matsuda invited me to observe several sessions of robot therapy at Leisure Time early in 2011. For this and several years after, these sessions tended to follow the same pattern. Members of the group set up their robots in a room on the first floor of Leisure Time. They then move to the second floor, where they pull several rectangular tables together to form two larger square tables. About a dozen residents, all in wheelchairs, are wheeled into the room by staff and positioned around the tables. Some who can still walk by themselves find their own places. As they come in, Okada-san announces their name in a voice loud enough for everyone to hear. Members answer with a hearty ' Konnnichiwa!' (good afternoon)! Once residents are seated around the tables, the activity begins.

On one day that I observed, I watched as the team placed robots on the tables in front of the seated residents, staggering about one robot per two residents. The robots included two robot cats (NeCoRo) and several AIBO robot dogs. As soon as the robots were placed on the table, the residents began to fiddle with them – touching them, petting them, smiling at them, or talking to them. Professors and students, who just a moment before had been intensely engaged in preparation, suddenly changed into amateur entertainers. They circulated among the seated residents, careful to crouch down to their level when talking with them. They helped facilitate the residents' verbal and tactile interactions with the robots, showing them how to get the robots to respond to touch or to voice. They functioned as prosthetic ears, repeating and interpreting the sounds coming from the robots when residents had trouble hearing them.

On the side of the room, I noticed another table with a laptop sitting on it. Two male undergraduates and their professor huddled around it, their eyes fixed on the screen. An AIBO robot sat next to the laptop. Its head moved in fits and starts. They set it upright and the robot took a few steps back and forth.

As I was trying to figure out what they were doing, a woman entered the room with her back bent at a nearly 90-degree angle. One of her arms hung down in front of her and the other was reaching under the waistband of her pants below her lower back, as if she were trying to tuck in her shirt. The professor and students near the laptop watched her carefully. She found a seat at a table. A few minutes later, one of the students walked over and placed the AIBO robot in front of her. He returned to the table and to the laptop. The professor left to join the others, but the students kept their eyes fixed on the woman.

In conversations with staff on previous visits to the nursing home, these students had learnt that the woman habitually roamed around the hallways of the nursing home. She pounded loudly on tables while sitting with other residents, ignoring the instructions of nursing home staff to stop, and occasionally entered the rooms of other residents uninvited. These are common symptoms of dementia, and nursing home staff interpreted them as expressions of stress, anxiety, or isolation. The students were curious to see if the frequency of these behaviours would change after one-on-one interactions with the robot. To gauge the effect, the students arrived in advance of the group to record the woman's movements in five-minute intervals. They repeated this protocol after the robot therapy session to measure any changes in the frequency or character of her wandering around the nursing home.[6]

Working with their professor, the students wrote a programme that would trigger movements in the AIBO robot (for example, moving forwards, moving backwards, sitting on its hind legs, wagging its tail, wiggling its ears, shaking its head or playing a song) on command. They could also turn on multi-coloured LEDs distributed along the robot's body in patterns or play pre-recorded sounds and phrases through the robot's speaker. Sony, the makers of the AIBO robot, originally designed the machine so such kinds of actions would emerge organically in the course of everyday interaction with a consumer 'owner'. In the context of a nursing home, where individual time with robots is limited, the cultivation of such intimacy is unrealistic. Members of the robot therapy team must supplement the capacity of the robot to sense and respond with their own ability to perceive and react based on experience and observation.

In choreographing the behaviour of the robot, they were aided by system software Sony engineers installed that allowed the robot to be remotely controlled via wifi. (They had set up a proxy wifi network in the room to make this possible.) Such direct control was not limited just to this experiment. The professors in the robot therapy group, all of them engineers, favoured the open operating system of robots like AIBO over those of robots like Paro and NeCoRo, which could not be remotely manipulated or to reprogrammed by anyone other than the manufacturer. On every one of my visits, there were always at least one or two AIBO robots that were controlled remotely. While the fur covering of a Paro or NeCoRo robot might invite an individual embrace, the open architecture of the AIBO robot facilitated collective interaction and engagement. It made possible the real-time orchestration of a form of controlled play that was at once more-than-human and more-than-robotic.

On the day I observed the session, this tight orchestration of the student experiment contrasted sharply with the thinly organised chaos around me. Machines whirred and whimpered. Songs and sounds blared out. Attention was attracted and then suddenly diverted. Conversations started and were suddenly interrupted. Robots ran out of juice and froze or got tripped up and fell over with a crash. Members of the group and nursing home staff shuffled around the room, alternately watching how residents were responding to the robots and keeping the machines from getting damaged in falls. They swept in to charge empty batteries and swap robots in and out to keep as many residents engaged as possible.

The relative immobility of the residents and the flurry of activity around them created a kind of theatre-in-the-round, with residents as critical audience, robots as actors, and team members as stage managers. It was a theatre of immediacy, one that generated feelings of enchantment and excitement through the spectacle of feeling machines. The audience was encouraged to observe, comment, and interact. Residents pointed at, laughed at, and talked about the robots as much as they tried to touch, push, pull, prod, hold, and pet them. Conversation was expected to break out: between residents and members of the team; in an imagined way between residents and robots; and, ideally, among residents themselves. Even when robots failed, they succeed. If they fell over or fell off the table, they became objects of laughter, robot actors doing robot slapstick.

As Okada-san remarked to me after one session concluded, being in nursing care places one in a near-constant state of dependence on others. This, he suggested, can be hard on a person's sense of dignity, especially when one is still conscious of declining cognitive ability. Interacting with an animaloid robot – an object that seems clearly inferior to you – can make you feel superior, more like an adult in control. Indeed, Okada-san stressed that the animal form of the robot is key. 'When it looks human', he told me, 'residents tend to stiffen up, thinking that they now have to deal with a person. The etiquette of the human world (*ningen kankei*) applies again. You can't relax and have fun. You must try to behave in line with ordinary rules of propriety.'

Animaloid robots here function like animals in animal therapy, but not exactly. In animal therapy, of course, it is animals, not robots, that are the central focus of activity. Animals attract interest not because they perform as something other than an animal; a dog attracts attention because it's a dog being a dog. In much the same way, the robots on the table-cum-stage play themselves. However, in contrast to animals, the behaviour of robots can be much more easily controlled. This can happen in real time, as in the case of the students directly manipulating the AIBO. It can also be programmed in advance. Either way, robots require the work of humans to perform much as humans need robots to create a spectacle beyond what humans can achieve alone. People are just too familiar; robots are not yet that nimble.

The aim of this structured and unstructured activity, members told me, is to create a shared basis for '*komyunikēshon*' (communication). Despite living together, residents of nursing homes rarely have much in common. They seldom share ties of family, community, occupation, or avocation. They come to the facility because of declining health, not for reasons of social affiliation. Their varying health statuses mean that even if a resident shares an interest with another resident, they might never recognise enough in common to engage in conversation. Living in a group without feeling a sense of community can lead to social isolation and hasten the progression of neurodegenerative disease.

Problems stemming from forced collectivity imply a collective solution. For this reason, the group makes sure that residents encounter robots together. (Even the individualised attention the woman I described above occurred at a table shared with other residents engaged with other robots.) At the centre of a theatrical round, robots become both objects of enchantment and conversation pieces for '*kyōtsū no wadai ga nai*' (people who lack common interests). Importantly, given the varying linguistic competence of residents with dementia, the content of the conversations that emerge is seen as less important than the very act of conversing in itself. The robot therapy group privileges this 'phatic function' of linguistic exchange – its capacity to affirm social ties without necessarily conferring meaning.[7] Residents might not understand what another resident says. Even if they do, the exchange might quickly be forgotten. Of greater importance is the moment of connection that these attempts at communication afford. These moments are believed to have therapeutic value insofar as they provide relief from feelings of dislocation and loneliness.

In sessions of robot therapy, then, the machines provide a kind of infrastructure for engagement. As an assemblage of material objects, the robot multitude mediates social interaction. By coming between persons, the robot multitude draws residents together with each other, with staff, and with members of the robot therapy group. But the mediation is not just physical. Devices like Aibo with open operating systems digitally mediate human–human interaction while masking the act of mediation itself. Remote-controlled AIBO robots move as if they were acting autonomously. Ultimately, then, the therapeutic effect of sessions depends surprisingly little on the capacity of any particular robot to sense or respond, of any particular feeling machine to feel. Instead, by providing a technological infrastructure for engagement, therapy robots become something other than (semi-)autonomous machines or media objects; they become platforms for sociality as care.

Conclusion: Robot Therapy as social infrastructure

In 2017 I returned to Japan for two months of follow-up research. I was fortunate to meet one day with Dr Matsuda and a younger colleague, Dr Inoguchi, who had started working with the robot therapy group a couple of years earlier. Over drinks one evening, we talked about how their approach had changed in the intervening years and how they saw their activities evolving into the future. Dr Matsuda shared his excitement about trying out some new robots in sessions but also lamented his group's failure to get more nursing home staff involved in robot therapy. The recognition of their difficulty in generating broader interest drew the conversation back to the reasons why members of the group found sessions of robot therapy valuable, if they were not readily replicated by nursing home staff themselves. I was struck by something Dr Inoguchi said in response, 'for me, I'm most interested in how "*chaneru*" [channels] of communication get created. I mean how it is that you can create the possibility of communication where no possibility existed before.' His comment about 'channels' of communication spoke directly to the phatic nature of sociality that I had earlier observed in the group's sessions at Leisure Time. But it also gestured towards something beyond the narrow field of dementia care itself. How best to create channels of communication is a problem that extends from engineering machines to facilitating human relationships writ large. Could there be any more pressing issue for a 'society of no-relation' than the fostering of phatic communion?[8]

The relevance of forging connections in a society of no-relation was likely on Dr Inoguchi's mind that night. Only a few hours before, we had concluded a meeting with a firm that had received a contract to promote a soon-to-open robot test field in Fukushima Prefecture. The test field was meant mainly as a site for robotics manufacturers to try out experimental rescue robots. But, given the devastation and fractured communities in the region following the triple disasters of March 2011, employees of the firm were curious whether robots could be used in the area for therapeutic purposes as well. Dr Matsuda described for them an idea that he thought might be workable which was clearly based on his work with the robot therapy group over the years. He suggested that they designate a central location near the test site as a robot therapy hub. From that hub, he told them, professional teams of robot therapists could travel out to conduct sessions at nursing homes and other locations, much like his group currently does. Care facilities, even families and smaller communities, could then receive the benefits of robot therapy without having to make the substantial investment of buying robots, maintaining and updating them, and training care staff to use them. Significantly, there was no suggestion that robot therapy would be limited to elderly individuals with dementia. The representatives spoke to a need for forging connection – 'channels' of communication – that extended throughout a region that had been coping with the effects of natural and nuclear disaster for the past several years. The meeting concluded with a clear expression of interest in this idea and a commitment to consider it further.

This meeting and Dr Inoguchi's comments after encapsulated for me the multiple meanings of robot therapy in Japan. On the one hand, members of the group have heeded the call of robot industry leaders to explore the possibilities of robotics in non-industrial contexts like entertainment and healthcare. They see themselves as tinkering with technologies of care, continually working toward more effective ways to work with robots in everyday life and in the particular context of dementia care. Sites of care provide for them a platform for experimentation through which their students can learn about the potential and limitations of robotics technology. On the other hand, their use of robotics technology to achieve care-through-sociality depends on the platformisation of robotics technology itself, in which humans supplement machine intelligence with their own expertise, perception, and dexterity. Humans and machines join together to create a space of interaction that is at once more-than-human and more-than-robotic. In its assemblage of humans and machines, the robot therapy group signals perhaps a way of maintaining bonds of affiliation in a society increasingly connected by digital technologies yet nevertheless plagued by isolation.

Notes

[1] These predictions were significantly revised in later MHLW documents (Wagner, 2010). See Arai (2009) for a critique of earlier projections.

[2] Translated directly, *kaigo robotto* means 'nursing care robot' in Japanese. There is no official definition of these robots (MHLW, 2014: 7) but the term generally refers to robots that are used in the long-term care of chronic or degenerative conditions. Since the phrase 'nursing care robot' is cumbersome in English, I use the shorter 'care robots' throughout this essay. There are reasons why this category of robot emerged in the first decades of the twenty-first century, which I do not have space to explore here.

[3] This name and references to all other individuals involved with the group are pseudonyms.

[4] Sony released AIBO in 1999 and regularly updated the robot until 2006.

[5] See Koch (2016) for a detailed analysis of these related terms.

[6] Here I am more interested in how the students conducted their experiment, rather than the results of the experiment itself. They did present the results of their intervention at a meeting of the group later in the year. They found that the controlled interaction with the robot decreased the expression of her symptoms only slightly.

[7] The 'phatic function' of language is one of linguist Roman Jakobson's six functions of language (Jakobson, 1990). Jakobson developed this notion from earlier work by the anthropologist Bronislaw Malinowski who called the social ties created through regular but not necessarily meaningful exchange 'phatic communion' (see Malinowski, 1936). Anthropologists have recently examined the infrastructural dimensions of phatic communication (Elyachar, 2010; Kockelman, 2010) as well as how the absence of phaticity can mark social disconnection (Nozawa, 2015).

[8] See Nozawa (2015).

References

Allison, A. (2013) *Precarious Japan*, Durham, NC: Duke University Press.

Allison, A. (2015) 'Discounted life: Social time in relationless Japan', *Boundary 2*, 42(3): 129–41.

Alzheimer's Association (2018) 'What is dementia?', Available from: https://www.alz.org/alzheimers-dementia/what-is-dementia [Accessed 20 December 2019].

Arai, H. (2009) 'Robotto sangyō ni kan suru shijō chōsa – shijō yosoku no hikaku to bunseki' [Comparison and analysis of market surveys and forecasts on robot industry], *Nihon robotto gakkai-shi*, 27(3): 265–7.

Brinton, M. (2011) *Lost in Transition: Youth, Work, and Instability in Postindustrial Japan*, Cambridge, UK: Cambridge University Press.

Brucksch, S. and Schultz, F. (2018) *Ageing in Japan – Domestic Healthcare Technologies: A Qualitative Interview Study on Care Robots, Monitoring Sensor Systems, and ICT-based Telehealth Systems*, Tokyo: German Institute for Japanese Studies.

Elyachar, J. (2010) 'Phatic labor, infrastructure, and the question of empowerment in Cairo', *American Ethnologist*, 37(3): 452–64.

Genda, Y. (2005) *A Nagging Sense of Job Insecurity: The New Reality Facing Japanese Youth* (J.C. Hoff, trans), Tokyo: International House of Japan.

Hardt, M. (1999) 'Affective Labor', *Boundary 2*, 26(2): 89–100.

Ishiguro, N.C. (2017) 'Robots in Japanese elderly care', in K. Christensen and D. Pilling (eds) *The Routledge Handbook of Social Care Work around the World*, London: Routledge, pp 256–70.

Jakobson, R. (1990) 'The speech event and the function of language', in L.R. Waugh and M. Monville-Burston (eds) *On Language*, Cambridge, MA: Harvard University Press, pp 69–79.

JARA and JMF (Japan Robot Association and Japan Machinery Federation) (2001) 'Nijūisseiki ni okeru robotto shakai-sōzō no tame no gijutsu-senryaku chōsa-hōkokusho' [Report on the technology strategy for creating a 'robot society' in the 21st Century], May, Tokyo: JARA.

Kitwood, T. (1990) 'The dialectics of dementia: With particular reference to Alzheimer's Disease', *Ageing and Society*, 10(2): 177–96.

Kitwood, T. (1997) *Dementia Reconsidered: The Person Comes First*, Philadelphia, PA: Open University Press.

Koch, G. (2016) 'Producing iyashi: Healing and labor in Tokyo's sex industry', *American Ethnologist*, 43(4): 704–16.

Kockelman, P. (2010) 'Enemies, parasites, and noise: How to take up residence in a system without becoming a term in it', *Journal of Linguistic Anthropology*, 20(2): 406–21.

Kosugi, R. (2008) *Escape from Work: Freelancing Youth and the Challenge to Corporate Japan* (R. Mouer, trans), Melbourne: Trans Pacific Press.

Lock, M.M. (2013) *The Alzheimer Conundrum: Entanglements of Dementia and Aging*, Princeton, NJ: Princeton University Press.

Lukács, G. (2015) 'The labor of cute: Net idols, cute culture, and the digital economy in contemporary Japan', *Positions*, 23(3): 487–513.

Lukács, G. (2020) *Invisibility by Design: Women and Labor in Japan's Digital Economy*, Durham, NC: Duke University Press.

Malinowski, B. (1936) 'The problem of meaning in primitive languages', in C.K. Odgen and A.I. Richards (eds) *The Meaning of Meaning*, New York: Harcourt, Brace, pp 296–336.

MHLW (Ministry of Health, Labor, and Welfare) (2014) *Fukushi-yōgu kaigo robotto kaihatsu no te-biki* [A Handbook on the Development of Care Robots and Welfare Technologies], Tokyo: MHLW.

Mol, A., Moser, I., and Pols, J. (2010) *Care in Practice: On Tinkering in Clinics, Homes and Farms*, Bielefeld: Transcript Verlag.

NHK Special (2010) *Muen Shakai – 'Muenshi' 32000-nin no Shōgeki* [Society of No Relation: The Shock of 32,000 'Lonely Deaths'], Tokyo: Nippon Hōsō Kyōkai.

Nozawa, S. (2015) 'Phatic traces: Sociality in contemporary Japan', *Anthropological Quarterly*, 88(2): 373–400.

Pew Research Center (2014) 'Attitudes about aging: A global perspective', Available from: https://www.pewresearch.org/global/2014/01/30/attitudes-about-aging-a-global-perspective/ [Accessed 15 December, 2019].

RAT/AAT Research Group (2002) 'RAT/AAT kenkyūkai seturitsu ni kan shi' [Regarding the establishment of the RAT/AAT Research Group], unpublished report, 28 January, Tokyo: RAT.

Robotic Care Devices Portal (2019) Available at http://robotcare.jp/jp/home/index.php [Accessed 10 December 2019].

Takeyama, A. (2016) *Staged Seduction: Selling Dreams in a Tokyo Host Club*, Redwood City, CA: Stanford University Press.

Tamura T., Yonemitsu, S., Itoh, A., Oikawa, D., Kawakami, A., Higashi, Y., Fujimooto, T., and Nakajima, K. (2004) 'Is an entertainment robot useful in the care of elderly people with severe dementia?', *The Journals of Gerontology: Series A*, 59(1): 83–5.

Wada, K., Shibata, T., Saito, T., and Tanie, K. (2004) 'Effects of robot-assisted activity for elderly people and nurses at a day service center', *Proceedings of the IEEE*, 91(11): 1780–8.

Wagner, C. (2010) '"Silver robots" and "robotic nurses"? Japanese robot culture and elderly care', in A. Schad-Seifert and S. Shimada (eds) *Demographic Change in Japan and the EU: Comparative Perspectives*, Düsseldorf: Düsseldorf University Press, pp 131–54.

Wright, J. (2018) 'Tactile care, mechanical hugs: Japanese caregivers and robotic lifting devices', *Asian Anthropology*, 17(1): 24–39.

Yamada, M. (2004) *Kibō kakusa shakai: 'makegumi' no zetsubōkan ga Nihon o hikisaku* [Hope Disparity Society: The sense of Despair among the 'Losers' Tears at Japan], Tokyo: Chikuma Shobō.

Yokoyama, K. (2004) 'Robotto shitī ni mukete no RT yunitto to RT konpōnento' [RT units and RT components in Robot City], *Nihon Robotto Gakkai-shi*, 22(7): 843–6.

Conclusion

Peichi Chung

This edited volume concludes with a discussion of how we answered five questions about technological change in media as they relate to work and play in Northeast Asia. The five questions are: (1) How does digital technology change labour practices and industry structure in electronic gaming? (2) How does play foster subjectivity in a corporation-dominated digital environment? (3) How do analogue and digital technologies afford meanings of work and play? (4) How do work and play in local settings challenge abstract concepts such as intellectual property, data privacy, sociality, and state-planned economy? (5) How are regions created through work and play of media contents and media ecosystems?

It does this by first establishing the theoretical context that explains the relevance of this book to the global technology industry and the transnational media flows that it helps create. It then uses this to organise the book's ten chapters around the five thematic questions that foreground interdisciplinary conjectures in the fields of political economy, cultural studies, game studies, and science and technology studies. This aids our subsequent attempt to explain our conceptualisation of technology, work, and play in the emerging techno-cultural spheres of Northeast Asia.

Here, we lay out the conceptual framework that theorises the daily practice of ICT use in the domains of work and play in Northeast Asia. We begin with our critical understanding of media technology that is comparable to the Western context. An urgent aspect emerges as we examine the consumption of media technology in Japan, South Korea, and North Korea. One major issue is the digital divide in media production and consumption among workers worldwide. Gray and Suri (2019) address the universal concern about class equality in their study of invisible platform workers in the US and India. Their research sheds light on the problem of the growing global underclass that competes as homeworkers who in turn contribute to the utopian dream of a platform economy crafted by the commercial narrative ideologically rooted in Silicon Valley. Similar concern also arises when we look at the issues of users' embodiment through various forms of media technology. Whether in the form of a DVD, USB, smartphone, location-based gaming technology, console, or arcade, media technology presents an underlining concern that equality and freedom for human development ought to be safeguarded as digital capitalism grows and merits the same degree of critical review of ICT for development in Northeast Asia.

The next aspect of this theoretical context points to new directions in the search to understand techno-regionalism in Asia. The focus on Japan and the two Koreas developed from an intention to bridge dialogue in the academic debate on transnational media flows. The phenomenon of media globalisation in East Asia is best illustrated by cases that describe the transnational media circulation of Japanese and Korean popular cultures. Iwabuchi (2002) has suggested that Japan's cute and 'odourless popular culture' has redefined the cultural imagery of Asian modernity through the global media consumption of Japanese media content. Jin (2016) reviews the cultural elements of transnationality in South Korean popular culture in the global social media landscape. Research on both of these popular cultures have revealed in them an interlinking relationship between Asia and the West. Both popular cultures tend to carry similar popular cultural forms and media aesthetics to Western popular culture as Japanese and South Korean cultural industries reshape local pop cultural elements to enter the global media distribution network. Progressing from this academic dialogue about transnational media flows, the book moves to a more complex level in order to make sense of the cultural linkages between Asia and the West through the representation of Northeast Asia's geopolitics.

As the book shows, Northeast Asian media's inter-regional cultural connections are based upon a longstanding contentious world politics shaped by the region's colonial past and post-war modernisation, which also shaped the development of the region's diverse media technology. After World War II, the region's three societies diverged along different developmental paths following the transformation of US, Chinese, and Northeast Asian political relations. Thus, the book provides insight into media representations of the region's geopolitics under the current ICT infrastructure that reflects the historical impact of Japan's colonial relationship to the Korean Peninsula, while also providing a closer angle from which to observe controversial media developments during and after the Cold War. It also offers an additional perspective on East Asia's transnational media flow in the light of its restructuring due to China's apparent decoupling from world politics during the early twenty-first-century phase of globalisation.

Specific chapters address theoretical concepts that further provide a specific understanding of the everyday practice of work and play in the region's various techno-cultural localities. The sequence of discussion includes concepts of subculture and resistance in the chapters by Jang, Galbraith, Amano and Rockwell; techno-surveillance in Marcén's chapter; technophobia in Sneep's chapter; corporeality and materiality, and piracy culture in Zhang and Lee's chapter; media hyperrealism and simulation in Shim's chapter; biopower and corporate governance in Bae's chapter; digital intimacy in Bender's chapter; and labour power in Chung's chapter. We use specific theoretical concepts to elaborate on the process by which cultural and media practices can cross the region's national boundaries.

The ten chapters are also grouped into three themes that demonstrate the various practices of work and play that can lead to different societal outcomes depending on local characteristics. In addition, the book discusses common inter-regional problems in the themes of gender and sex online, governance and regulations, as well as techno-identity and digital labour condition. Specifically, in the context of work and play, the chapters by Jang and Galbraith demonstrate how technology enables a gendered techno-cultural space to legitimate subcultural and women's experiences through play. The chapters by Marcén, Sneep, Lee and Zhang, and Shim explain the extent to which media power triggers fear and nationalist sentiment regarding the capacity of state regulation. The chapters by Bae, Bender, Amano and Rockwell, and Chung collected in the last theme, techno-identity and digital labour conditions, speaks about how digital technology enables the cultural re-imagination of a nation, an ethnic group or a social group through play.

We further answer the following five questions by explaining how each chapter presents the practice of work and play in various techno-cultural spheres of Japan and the two Koreas.

The changing labour practices and industry structure in electronic gaming

Reviewing our examination of the changing corporation–labour relationship through gameplay, we find that that platforms have encouraged a new type of labour practice in electronic gaming. In this context, one of the most concerning issues in digital labour has been the unrewarded labour practices that persist in the digital economy (Andrejevic, 2013). Having documented playful experiences on a variety of platforms, including adult computer games, *Pokémon GO*, pachinko, *Overwatch* and *League of Legends*, we discern a platform-based labour practice that involves the ability to operate computer machines. The skills involved in this new type of labour practice in electronic gaming surpass the skills needed to undertake paid labour in factory operation. Indeed, this unpaid labour practice requires gaming expertise based upon one's knowledge of how to perform specific activities, control data flows and access high-speed and networked computers – not to mention the computational thinking required in video gameplay, as stressed by Sicart (2014). Labour practices in electronic gaming also involve computational thinking in order to engage timely human-to-system productivity. Similar cases of human-to-machine capabilities are shown in the chapters by Bae and Bender. For instance, esports players develop bodily capabilities such as biopower in order to game competitively. This new development also represents a change in the work environment regarding value creation in the emerging platform-based economy. Elder-care labour through robotics as in the case of Japan's robot therapy also shows the change of labour practices in processes of robotic automation.

At the industry level, the growing influence of the game corporation is one of the most prevalent developments determining the global electronic gaming industry's current concentrated structure (Kerr, 2016). As game corporations develop entertaining titles in hopes of luring gamers to become immersed in the virtual gaming world, work resources in electronic gaming outside the commercial activities of the corporation remain limited. In East Asian societies, electronic gaming also suffers from negative images of addiction and gambling, and this creates a lack of trust in game corporations in the light of their growing influence in the society. Bae's chapter notes that the global popularity promoted by game corporations offers a form of governance by game developers. We often see conglomerates support electronic gaming in the form of large monetary investments in support of esports tournament and the cultivation of greater esports enthusiasm. In the case of *Pokémon GO*, Sneep mentions another social phenomenon – phobias in Japanese society – through young gamers seen playing augmented reality games at dangerous places in Japanese cities. The blurring of the boundary between the real and virtual world leads players to undermine public safety even as they more easily enter the story of *Pokémon GO*. Chung also mentions the importance of implementing public welfare support to improve South Korea's overtly commercial driven esports ecosystem. All the chapters concerning issues of corporation and labour relationships collectively point to a potential problem of trust if corporatisation and monetisation continue to be the central issue of concern in electronic gaming's global industry structure.

Fostering subjectivity through play in a corporation-dominated digital environment

The chapters by Galbraith and Sneep showed that play has become a powerful mode of expression that fosters imaginative spaces for subcultural identities. Sicart (2014) notes that it is important to remain playful and pleasurable when one acts as an agent in the virtual environment. Play in a corporation-dominated digital environment requires performance in a theatrical setting of the corporation's creation. The environment differs in every game so that play can be contextual. Play can also be creative as players type texts, read screens, and act freely to enjoy their playtime. However, play sometimes can also be disruptive to social hierarchies. The open environment in a game invites players to interact to foster sociality in the game space. In particular, when gamers hold protests in computer games, play becomes a collective form of political resistance and digital activism. In all, there is no right way to play a game. Gaming is like theatre in that it allows players to perform differently every time. In this way, play contains a disruptive aspect.

Open digital environments symbolise democratic social spaces in which players can shape subjectivities and express their minds. In highly competitive societies like Japan and South Korea, the digital environment offers an alternative space where people can cultivate subcultural identities. Galbraith's chapter examines adult computer games; he observes how its subculture, replicated in cultural discourse as sexual desire in anime and manga characters, manages to make enough noise to affirm its presence in Japan's nationalist discourse. The techno-cultural space that adult computer games occupy proves the power of a subculture to manoeuvring its way into participation in the nationalisation process for defining 'Cool Japan'. Galbraith also takes note of the politics involved in regulating the adult computer game industry. The strong nationalist sentiment in Japan articulates an exclusive discourse of 'Japan Only' in rebranding this as 'Cool Japan', which was designed to attract foreign markets. Such exclusivity is challenged at the industrial level as adult computer game producers maintain their creative right of production to push boundaries in order to challenge 'Japan Only' national branding discourse. Importantly, the power of this revolt from a subcultural position requires favour and support from fans in the global community.

Jang's study of female entrepreneurship in South Korea's social media examines strategies that female social media entrepreneurs develop in order to promote publicity for their e-commerce businesses. The chapter explores how work in the digital economy remains a feminised concept in which women entrepreneurs are seen presenting traditional images of gender roles to create a visual profile of their online presence. The society's animosity acknowledged in Sneep's chapter depicts the same issue from the opposite side. That is, despite *Pokémon GO's* ability as an innovative augmented reality game to offer a new sense of play in the city, there persists the public assumption that smartphone use is an unhealthy addiction. This backlash against virtual reality gaming reflects a larger societal context, as gaming culture extends itself to challenge still dominant traditional social values in the digital space.

Fostering meaning through play and work with analogue and digital technologies

Both analogue and digital technologies are platforms that are equally effective at developing meaning in work and play. In the analogue media world of North Korea, film and television are the dominant media forms that shape the hyperreal national image of the country. Shim's chapter scrutinises the media works that present North Korea as a post-famine nuclear nation. The images presented to the global audience are propagandistic and militaristic, not to mention hyperreal, however influential they might be in world politics. Conceptually grounded in Baudrillard's 'hyperrealism of simulation', Shim describes North Korea's manipulation of media representation in order to engage foreign diplomacy through simulated nuclear weaponry. The country controls sign production to create media spectacles that grant it the power to negotiate with global powers in the digital media space. North Korea's soft power initiative shows how the imaginations of North Koreans have been shaped through the country's manipulation of propagandistic film and news reports from Western media. Baudrillard argues that the 'hyperreal is beyond representation' (1993: 75). Because simulated media space can contaminate reality, representative signs in global media flows can reproduce meanings to operate at the symbolic level and redefine media reality.

In contrast to analogue technological space, the mediated world interconnected through underground and unofficial channels tells contradictory stories in North Korea. To explain, Lee and Zhang explore in their chapter the meaning-making process in piracy culture. VCR videos, DVDs, and USBs constitute the dominant source of learning for North Korean consumers who wish to gain knowledge about the outside world. The authors explain the physical dimensions of cross-border media circulation whose flows rely upon North Korea. Smugglers are part of the trusted channel for circulating Western media content into the country. Under the authoritarian state, fans of Korean Wave and Hollywood enjoy their viewing pleasure by becoming a part of the audience community outside North Korea. As the reception of capitalist and democratic values occurs through the process of viewing pirated media, North Koreans who left the country as defectors reveal the power of pleasurable viewing.

Challenging abstract concepts such as intellectual property, data privacy, sociality, and a state-planned economy

The book covers several issues relating to media laws and state-planned economies. Lee and Zhang offer legal discussion of media circulation under the regime of intellectual property rights. It provides a comprehensive analysis that exposes the complex meaning-making process in the development of media flows that operate outside the regulatory framework of intellectual property. The chapter shows that media technology opens up cultural space for people to imagine life and humanity beyond national boundaries. As the concept of intellectual property rights is utilised to control media cultural flows through taxes, tariffs, and trade sanctions, digital technology helps to extend the geographical reach of American and South Korean soft power. It might be crucial to reconsider the ability of soft power to influence the minds of North Korean media fans outside global intellectual property regimes.

Similarly, Marcén's chapter seeks a regulatory model of data governance. Her chapter on data privacy laws in Japan and South Korea raises awareness about rights that should be safeguarded in the domain of digital data transmission. Digital technology offers advanced speeds of communication. Should data be regulated so as to satisfy surveillance purposes in Northeast Asia? Do corporations and states also own the rights to personal data? What is the national characteristic when the concept of privacy is regulated in Japan and South Korea?

Answers to these questions have indicated an important area that influences how sociality takes place during play and work in the digital communication environment of Northeast Asia.

The creation of regions via media contents and ecosystems

Finally, the book engages in the conceptualisation of region in the domain of media-forming powers. Chapters in this book present various regionalist imaginaries of Asia created through media content and ecosystems in Northeast Asia. The elaboration of Asianism sheds light on a contested narrative that this book aims to construct in Asian imagination beyond 'the Orient', 'East Asia', and 'Southeast Asia'. In the journal article 'Globalizing the regional, regionalizing the global: Mass culture and Asianism in the age of late capital', Ching (2000) sees understanding the complexity of regional (anti)-Japanese colonial narratives and sentiments as important work to articulate Asia's relationship with Western imperialist power. According to Ching, as region becomes an effective concept to realise the internal self-efficient system that rejects Western universalism, this book extends Ching's thesis to formulate Asia's newly evolving regional identity by constructing the region's techno-reality shaped by media contents and ecosystems of Japan and the Two Koreas.

Steinberg and Li (2017) propose the idea of 'regional platform' to explain a new type of digital media flow that is made available by Asian platforms such as Korea's KakaoTalk, Japan's Niconico Videos and China's WeChat and Bilibili. A similar line of thought on media-forming powers in Northeast Asian media space is also prevalent in this book. The chapters by Galbraith, Zhang and Lee, and Bae directly describe how regionalist imaginaries about Northeast Asia are formed through certain media reactions. Galbraith's chapter shows a unique regionalist imaginary in Japan's fascination in adult computer game. Zhang and Lee's chapter frames a regionalist imaginary by examining how Hollywood and South Korean intellectual property regimes are enacted in relation to bodies in the North Korean context. Finally, Bae presents a case of regionalist imaginary of biopolitical bodies as media reports the presence of a large number of South Korean players in North American esports teams.

We hope the chapters in this volume have demonstrated that media technology practices for work and play in Northeast Asia are not exceptions to those of the West. At the same time, the governments, corporations, and audiences in the three Northeast countries are not blindly trying to catch up with the West. We have shown that Westernisation, colonisation, independence, and post-war modernisation have provided political, economic, and social conditions for divergent media technological developments in the three countries. What is yet to be seen is how the fast-changing power redistribution between the US and China will influence the ecology of media technologies for work and play in Northeast Asia.

References

Andrejevic, M. (2013) 'Estranged free labour', in T. Scholz (ed) *Digital Labour: The Internet as Playground and Factory*, New York: Routledge, pp 149–64.

Baudrillard, J. (1993) *Symbolic Exchange and Death*, London: Sage Publications.

Ching, L. (2000) 'Globalizing the region, regionalizing the global: Mass culture and Asianism in the age of late capital', *Public Culture*, 12(1): 233–57.

Gray, M. and Suri, S. (2019) *Ghost Work: How to Stop Silicon Valley from Building a New Global Underclass*, Boston, MA: Houghton Mifflin Harcourt.

Iwabuchi, K. (2002) *Recentering Globalization: Popular Culture and Japanese Transnationalism*, Durham, NC: Duke University Press.

Jin, D. (2016) *New Korean Wave: Transnational Cultural Power in the Age of Social Media*, Champaign: University of Illinois Press.

Kerr, A. (2016) *Global Games: Production, Circulation and Policy in the Networked Era*, New York: Routledge.

Sicart, M. (2014) *Play Matters*, Cambridge, MA: The MIT Press.

Steinberg, M. and Li, J. (2017) 'Introduction: regional platforms', *Asiascape: Digital Asia*, 4(3): 173–83.

Index